AN INTRODUCTION TO

THE ALGAE

Ian Morris

Lecturer in Botany
University College London

HUTCHINSON UNIVERSITY LIBRARY

LONDON

HUTCHINSON & CO (*Publishers*) **LTD**
178–202 Great Portland Street, London W1

London Melbourne Sydney
Auckland Bombay Toronto
Johannesburg New York

★

First published 1967
Second edition 1968

*This book has been set in Times, printed in Great Britain
on Smooth Wove paper by Anchor Press, and
bound by Wm. Brendon, both of Tiptree, Essex*
09 080712 x (cased)
09 080713 8 (paper)

CONTENTS

ACKNOWLEDGMENTS

I am deeply indebted to four eminent phycologists, Professor G. E. Fogg, Dr M. Parke, Dr P. S. Dixon and Professor M. B. E. Godward, for reading the manuscript and for making numerous invaluable suggestions. I am also grateful to Dr D. P. Kelly for a preliminary reading of the manuscript, and to Dr W. G. Chaloner for reading parts of the manuscript (notably Chapter 14) and for many helpful discussions on evolutionary topics. Finally, I wish to thank the Secretaries in the Botany Department of University College, London for typing the manuscript.

PREFACE

The writing of a small book on a large subject is an unenviable task. When the subject is as large as the present one, a small book such as this is in danger of merely presenting meaningless generalisations and erroneous oversimplifications. The present book has been written with this danger clearly recognised, but it has also been written with the conviction that such a book can fulfil a useful function.

Clearly it is not possible to consider all aspects of the biology of algae. Instead the book confines itself to a discussion of the structure and reproduction of algae; unfortunately, subjects such as their biochemistry, physiology and ecology have had to be omitted (fortunately these features are emphasised in the recent larger book of Round [209]). Moreover, when discussing the structure and reproduction of algae, it has been necessary to limit the remarks to those relevant to the general point under discussion at the time, and consequently no detailed picture of any particular species or genus has been given. Similarly, space has not permitted a description of the habitats, geographical distribution, etc., of any alga which is mentioned. It is therefore hoped that readers will treat this book as a brief supplement to the larger standard texts, where many of the omitted details may be found.

Within the limits imposed by the above restrictions, the book has been written with two main aims. First, to select from the wealth of detail of many of the larger groups of algae those features which appear to reflect the fundamental developments of morphology and reproduction. Secondly, to emphasise the way in which modern techniques of electron microscopy, comparative biochemistry, etc., are increasing our understanding of algae, particularly of many of the smaller phyla.

This latter aim appears to be important since students too frequently assume that the structure, reproduction and classification of organisms are so well established that the need for future investigations no longer exists. This is clearly not so for any group of organisms; and is particularly not so for the Algae. Because importance has been attached to recent investigations of some of the smaller algal phyla, the space devoted to them, in proportion to the larger groups, is greater than is found in most books. Thus, any tendency towards generalisation and over-simplification will be particularly prominent in the chapters on the larger groups, notably the Chlorophyceae, Rhodophyceae and Phaeophyceae. Despite this, it is to be hoped that the essential features of those classes have been summarised.

For the second edition I have taken the opportunity of having the illustrations redrawn and, in some cases, rearranged and slightly altered.

I

INTRODUCTION AND GENERAL
CLASSIFICATION OF THE ALGAE

The Algae comprise a large heterogeneous assemblage of plants, and characteristics by which they can be defined are difficult to specify. The most comprehensive definition of the group appears to be that of Fritsch,[90] who says: 'Unless purely artificial limits are drawn, the designation alga must include all holophytic organisms (as well as their numerous colourless derivatives) that fail to reach the level of differentiation characteristic of archegoniate plants.'

When deciding whether a particular plant has reached the 'level of differentiation characteristic of archegoniate plants', the particular feature usually emphasised is that the sex organs and sporangia of Bryophyta and all plants above them in the evolutionary scale are multicellular with the outer cells sterile, whereas in algae all sporangia and sex organs are either unicellular, or when multicellular, all cells are fertile. However, the reproductive structures of the Charophyceae (see pp. 74–7) are apparently multicellular; and although the sterile tissue is generally interpreted as being vegetative, and not part of the reproductive structure, this interpretation (and consequently the above distinction between algae and archegoniate plants) is doubtful.

An alternative method of distinguishing between them is to emphasise the uniform structure of the archegonium in archegoniate plants and to recognise that there is no such structure in algae.

ORGANISMS TO BE INCLUDED IN THE ALGAE

Although most authorities now accept the definition proposed by Fritsch, earlier attempts to define algae used more positive features, and the group was therefore more restricted than at present. Thus, at the end of the nineteenth century algae were thought to include four classes, Myxophyceae, Chlorophyceae, Phaeophyceae and Rhodophyceae (blue-green, green, brown and red algae respectively), together with the diatoms (Bacillario-phyceae) which were included, either as a separate group, or as part of the Phaeophyceae. Motile unicells and colonial forms were not included, since they were classified as members of the Flagellata in the Protozoa (the *Chlamydomonas—Volvox* series was an exception). The cells of flagellates were thought not to have a wall, to readily adopt a resting stage, to divide only in the longitudinal plane and not to show sexual reproduction. The cells of algae were usually thought to have a wall, not to adopt resting stages, to divide in both the longitudinal and transverse planes and to reproduce sexually.

As an increasing number of organisms was examined it was found that many 'flagellates' were closely related to 'algae' and the distinction between many of them was seen to be artificial. Moreover, when related 'algal' types and 'flagellate' types were put in the same taxonomic group, it was possible to recognise several classes showing striking parallel changes in morphological and functional complexity. Such parallel lines of development are generally assumed to indicate that the arrangement of organisms reflects their supposed evolutionary relationships.

THE PRIMARY CLASSIFICATION OF ALGAE

In this book algae are classified into 10 divisions: Cyanophyta, Chlorophyta, Xanthophyta, Chrysophyta, Bacillariophyta, Pyrrophyta, Cryptophyta, Euglenophyta, Phaeophyta and Rhodophyta. This chapter is concerned solely with some of the criteria used to classify algae into these divisions; the further classification into classes, orders, families, genera, etc., is not considered at this stage, but is discussed later in the book when each of the divisions is described separately. Two small classes—Chloromonadophyceae and Prasinophyceae—are also described (in Chapter 11) but are not included in phyla since much is still unknown.

In general, details of vegetative structure and processes of reproduction are not particularly useful for the primary classification of algae (of course such details are important for algal classification at the level of species, genus, family, etc.). Instead, the primary classification of algae is based on five main criteria of a different nature: 1. photosynthetic pigments, 2. the nature of the food reserves, 3. the nature of cell wall components, 4. the types of flagella, and 5. certain details of cell structure. Each of these criteria is described separately below, but it must be emphasised that the final classification of algae depends on a combination of *several* characters, and not on any single feature.

1. PHOTOSYNTHETIC PIGMENTS[234]

Algae from the various phyla show striking differences of colour and these often afford a quick guide to a preliminary classification of an alga. However, the colour frequently varies with changes in environmental conditions and an accurate classification depends on chemical analyses of the photosynthetic pigments.

There are three kinds of photosynthetic pigments in algae—chlorophylls, carotenoids and biloproteins—and the distribution of these is important in algal classification.

1. *Chlorophylls*[12]

Chlorophylls extracted from different algae show different spectral properties; on this basis a number of different chlorophylls have been recognised, and have been termed chlorophylls *a*, *b*, *c*, *d* and *e*.

The distribution of these chlorophylls amongst various algal groups is shown in table 1. Chlorophyll *a* is present in all algae, as it is in all photosynthetic organisms other than photosynthetic bacteria. Chlorophyll *b*, the other chlorophyll of higher plants, is found in the Euglenophyta and Chlorophyta, and in no other algal division. Chlorophyll *c* is the most widespread of other chlorophylls, being present in members of Bacillariophyta, Chrysophyta (?), Pyrrophyta, Cryptophyta and Phaeophyta. Chlorophyll *d* appears to be present only in the Rhodophyta and chlorophyll *e* has been identified in only two genera of Xanthophyta, *Tribonema* and the zoospores of *Vaucheria*.

2. *Carotenoids*[102, 178]

Carotenoids are of two kinds, carotenes and xanthophylls. Carotenes are linear unsaturated hydrocarbons and xanthophylls are oxygenated derivatives of these. β-carotene is present in most algae, although it is replaced by α-carotene in the Caulerpales of the Chlorophyceae, in the Cryptophyta and to a lesser extent in the Rhodophyta (see table 1). In the Charophyceae (a class within

TABLE 1

*The distribution of photosynthetic pigments in algae**

	Cyanophyta	Chlorophyta	Xanthophyta	Chrysophyta	Bacillariophyta	Pyrrophyta	Cryptophyta	Euglenophyta	Phaeophyta	Rhodophyta
Chlorophylls										
chlorophyll *a*	+	+	+	+	+	+	+	+	+	+
chlorophyll *b*	—	+	—	±	—	—	—	+	—	—
chlorophyll *c*	—	—	—	?	+	+	+	—	+	—
chlorophyll *d*	—	—	—	?	—	—	—	—	—	±
chlorophyll *e*	—	—	+	—	—	—	—	—	—	—
Carotenoids										
Carotenes										
α-carotene	—	+a	—	—	—	—	+	—	—	+
β-carotene	+	+	+	+	+	+	—	+	+	+
γ-carotene	—	+b	—	—	—	—	—	—	—	—
lycopene	—	+b	—	—	—	—	—	—	—	—
ε-carotene	—	—	—	—	+	—	+	—	—	—
Xanthophylls										
lutein	+?	+	—	+?	—	—	—	—	+	+
violaxanthin	—	+	+?	—	—	—	—	—	+	+?
fucoxanthin	—	—	—	+	+	—	—	—	+	—
neoxanthin	—	+	+?	—	—	—	—	+	—	—
astaxanthin	—	+	—	—	—	—	—	+	—	—
diatoxanthin	—	—	—	+	+	—	—	—	+	—
diadinoxanthin	—	—	—	+	+	+	—	—	—	—
zeaxanthin	+	+b	—	—	—	—	+?	—	—	+
peridinin	—	—	—	—	—	+	—	—	—	—
dinoxanthin	—	—	—	—	—	+	—	—	—	—
taraxanthin	—	—	—	—	—	—	—	—	—	+
antheraxanthin	—	—	—	—	—	—	—	+	—	—
myxoxanthin	+	—	—	—	—	—	—	—	—	—
myxoxanthophyll	+	—	—	—	—	—	—	—	—	—
oscilloxanthin	+	—	—	—	—	—	—	—	—	—
echinenone	+	—	—	—	—	—	—	+	—	—
Biloproteins										
C-phycocyanin	+	—	—	—	—	—	—	—	—	—
C-phycoerythrin	+	—	—	—	—	—	—	—	—	—
R-phycocyanin	—	—	—	—	—	—	—	—	—	+
R-phycoerythrin	—	—	—	—	—	—	—	—	—	+
Phycocyanin ⎫ unknown	—	—	—	—	—	—	+	—	—	—
Phycoerythrin ⎭ types	—	—	—	—	—	—	+	—	—	—

* The data in the above table are taken from refs. 12, 102, 178 and 182

a α-carotene is present in the Caulerpales

b γ-carotene, lycopene and zeaxanthin are present in the Charophyceae

the Chlorophyta) β-carotene is replaced by two carotenes which are characteristic of photosynthetic bacteria, lycopene and γ-carotene. Another carotene, ϵ-carotene, is characteristic of Bacillariophyta and has also been identified in some members of the Cryptophyta.

There are many different xanthophylls in algae, and since many are unique to particular algal groups, they are important diagnostic features (table 1). For example, peridin is found only in the Pyrrophyta, myxoxanthin and others only in the Cyanophyta, taraxanthin in the Rhodophyta and antheraxanthin in Euglenophyta.

3. *Biloproteins*[182]

Chlorophylls and carotenoids are soluble in lipid solvents and cannot be extracted in aqueous solution. However, water-soluble pigments can be extracted from some types of algae. During the extraction procedure the free pigment cannot be separated from a protein moiety and the name of the pigments was therefore changed from *phycobilins* to *biloproteins* to indicate the existence of the pigment-protein complex. Biloproteins are present in only three algal divisions, the Cyanophyta, Rhodophyta and Cryptophyta (table 1), (see also *Cyanidium*—chapter 11). Analysis of the spectral properties of these pigments shows that there are two kinds of biloproteins, phycocyanin and phycoerythrin. Moreover, each of these shows differences between the three groups. In general, those of the Cyanophyta are of the C-type, those of the Rhodophyta are of the R-type and those of the Cryptophyta are of a third kind.

The proportion of one kind of pigment to the other is variable. For example, cells of the Chlorophyta and Euglenophyta appear green because of an excess of chlorophylls over carotenoids, whereas the yellow-brown colour of groups such as the Bacillariophyta, Chrysophyta, Pyrrophyta, Cryptophyta and Phaeophyta and the yellow-green colour of the Xanthophyta reflect an excess of carotenoids compared with chlorophylls. Also the characteristic colour of the Cyanophyta (blue-green) and the Rhodophyta (red) are due to an excess of the appropriate biloproteins. However, the proportion of one type of pigment to the other can vary considerably with changes in the environmental conditions,[82, 83] and it is difficult to justify its use as a taxonomic feature.

2. FOOD STORAGE PRODUCTS[171, 174]

The initial stages in carbon dioxide fixation are probably the same in all photosynthetic organisms. Thus, the primary products of photosynthesis are the same in all algae. However, the *insoluble* products which accumulate over a longer period of time are more variable and they afford useful taxonomic criteria. The compounds which are most widespread and most useful in the primary classification of algae are various polysaccharides. 'True' starch, similar to that found in higher plants, is only found in one algal division, the Chlorophyta. Two other divisions, the Rhodophyta and Cyanophyta, accumulate characteristic starches; these are *floridean starch* and *myxophycean starch* respectively. Both are polyglucose molecules identical to the amylopectin part of higher plant starch. 'True' starch, floridean starch and myxophycean starch are molecules in which the glucose sub-units are joined by α-1,4 linkages (the contrast between α-1,4 and β-1,3 linkages is shown in fig. 1). Three polysaccharides with β-1,3 linkages are now recognised as being important constituents of algae.

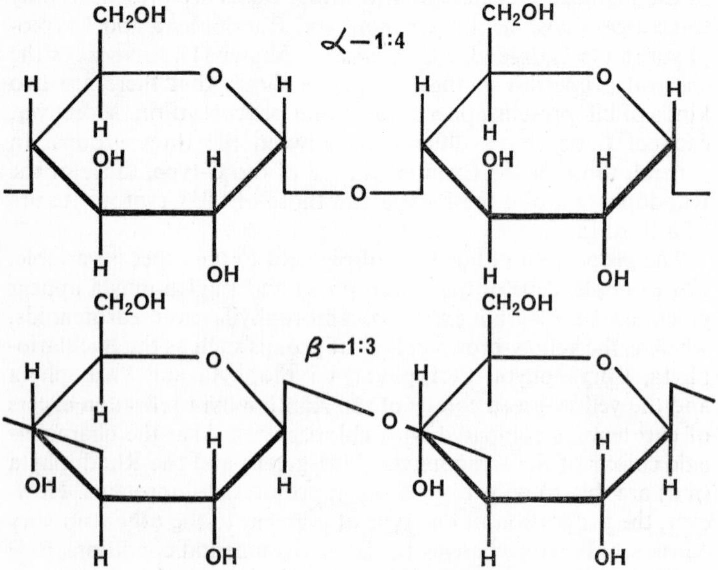

Fig. 1

α 1-4 and β 1-3 linkages of algal polysaccharides

These are *laminarin* (Phaeophyta), *paramylum* (Euglenophyta) and *chrysolaminarin*, or *leucosin* (Chrysophyta and Bacillariophyta).

Other storage products of taxonomic importance include *floridoside* and *mannoglycerate* in the Rhodophyta, and the proteinaceous *cyanophycin* found in cells of blue-green algae (Cyanophyta). *Mannitol*, a polyhydroxy alcohol, was thought to be found only in the Phaeophyta but it has now been identified in some red algae (Rhodophyta), and in some dinoflagellates.[10]

Fats accumulate in a large number of algae, and the proportion of fat to other storage products varies considerably. For example, cells of the Chlorophyta usually accumulate more carbohydrate than fat whereas cells of Chrysophyta and Bacillariophyta accumulate a high proportion of fat. However, the proportion of fat varies with changes in environmental conditions[82, 83] and its validity as a taxonomic feature is difficult to justify. A more promising approach to the importance of fats in algal classification appears to centre on their chemical nature. Thus, from the limited evidence available it appears that the degree of unsaturation can be correlated with certain algal divisions.[174] Fats of green algae (Chlorophyta) resemble those of higher plants, whereas those of some Phaeophyta are less unsaturated and those of some Rhodophyta are more unsaturated.

Sterols have been thought to be of importance in the primary classification of algae,[23] although the evidence is extremely limited. Cyanophyta (in common with bacteria) differ from all other organisms in lacking sterols.[142] A general picture of the distribution of sterols amongst the other algal divisions is shown in table 2. However, the situation is not as clear-cut as suggested by that table. Although sitosterol, the sterol most abundant in higher plants, is widespread in the Chlorophyta some anomalous results have been obtained; for example, cells of the green alga *Chlorella pyrenoidosa* accumulate ergosterol, the major sterol of fungi. Similarly, although fucosterol is the most common sterol of the Phaeophyta, it is replaced by the closely related sargasterol in some species, and it has also been identified in some species of Chlorophyta, Rhodophyta, Chrysophyta and Bacillariophyta. Cholesterol appears to be present in all Japanese species of red algae, whereas British species contain either sitosterol or fucosterol. The explanation for this is unknown, but if it reflects a general tendency for the sterol content to vary with changes of environmental conditions, the use of sterols as taxonomic 'markers' would be difficult to justify.

16   An Introduction to the Algae

TABLE 2

Distribution of sterols in algae*

	Cyanophyta	Chlorophyta	Xanthophyta	Chrysophyta	Bacillariophyta	Pyrrophyta	Cryptophyta	Euglenophyta	Phaeophyta	Rhodophyta
Sitosterol	—	+	+	—	—	?c	?c	—	—	+
Fucosterol	—	—a	—	+	+	?	?	—	+	+
Sargasterol	—	—	—	—	—	?	?	—	+d	—
Cholesterol	—	—	—	—	—	?	?	—	—	+e
Ergosterol	—	—	—	+	—	?	?	+	—	—
Chondrillasterol	—	—b	—	—	+	?	?	—	—	—

* The data are taken from the paper of Miller[174]
a Fucosterol has been identified in a species of *Cladophora*
b Chondrillasterol has been found in a species of *Scenedesmus*
c Pyrrophyta and Cryptophyta do not appear to have been examined
d Sargasterol is much less common than fucosterol
e Cholesterol is sometimes replaced by dehydrocholesterol

3. CELL WALL COMPONENTS

When a wall is present, its chemical constituents vary from one group to another and are sometimes important indications of the taxonomic position of a particular alga. The cell wall is generally made up of two sorts of material, an inner water-insoluble material, and an outer pectic or mucilaginous substance soluble in boiling water.[133] Although both inner and outer materials are polysaccharides, lipid and proteinaceous materials are also present. The most common water-insoluble polysaccharide of the inner layers is cellulose, and this is present in walled species of all divisions except the Chrysophyta and Bacillariophyta (there is also some doubt about the Cryptophyta and the Cyanophyta). X-ray diffraction patterns show the existence of different kinds of cellulose, depending on the particular crystalline arrangement. As yet, insufficient data are available to indicate whether such differences would support the existing primary classification of algae, although the details are already being interpreted as indicating certain relationships between algae within a particular division (e.g. see p. 67).

Other characteristic components of the cell walls include the polyuronic acid, *alginic acid*, found in the walls of Phaeophyta, *fucinic acid*, also found in some of the larger species of Phaeophyta and a characteristic *mucopeptide component* present in the cell walls of blue-green algae (Cyanophyta) (see p. 27). Certain algae, particularly the Chrysophyta and Bacillariophyta, have a marked tendency to have silicified walls.

4. FLAGELLA

Apart from the Cyanophyta and Rhodophyta, flagella are found in all other divisions of algae, and their nature, number and position are important characters for the primary classification of the Algae. The detailed fibrillar structure of algal flagella resembles that of cilia and flagella of other organisms (except bacteria) in showing the typical '9+2' pattern of component fibrils.[15, 154] The macrostructure of algal flagella does not, however, show such uniformity. For a long time algal flagella were thought to be of two kinds: *acronematic* ('whiplash', smooth, peitschengeisel), and *pantonematic* ('flimmer', flimmergeisel). The former, as the name implies, are smooth and whip-like whereas the latter have longitudinal rows of fine hairs arranged along the axis of the flagellum. More recent work with the electron microscope has revealed at least one other kind in which the flagellar surface is covered by minute hairs (different from those on the pantonematic flagella) and scales. At the moment algae with such flagella are included in a new class, the Prasinophyceae (see p. 130). A brief outline of the use of flagella in algal classification is given in table 3; a more detailed discussion of the connexion between flagella and classification is given in the recent review of Manton.[158]

5. CERTAIN ASPECTS OF CELL STRUCTURE

It is possible to identify some divisions of algae from certain details of their cell structure, and the more important cell features by which such algae are characterised are outlined in table 3. Cells of blue-green algae (Cyanophyta) are *procaryotic*, whereas those of all other algae (also all other cellular organisms other than bacteria) are *eucaryotic*. Stanier and van Niel[225] have summarised the features of procaryotic cells as follows:

(a) The absence of internal membranes which separate the resting nucleus from the cytoplasm, and which isolate the enzymatic machinery of photosynthesis and respiration in specific organelles (that is, no mitochondria or chloroplasts can be identified).

TABLE 3

*Types of flagella in algae and details of cell structure
used in the primary classification of algae*

	Flagella	*Details of cell structure*
Cyanophyta	none	Procaryotic cells
Chlorophyta	2 or 4 anterior equal, acronematic	
Xanthophyta	2 unequal, anterior 1 acronematic, 1 pantonematic	
Chrysophyta	acronematic and pantonematic (+haptonema)*	in some, cell surface is covered by characteristic scales *
Bacillariophyta	1 pantonematic— anterior	cell in two halves, the walls silicified with elaborate 'markings'
Pyrrophyta	1 acronematic 1 'band-shaped'*	in most, there is a longi- tudinal and transverse furrow and angular plates
Cryptophyta	2 equal, lateral pantonematic	gullet is present in some species
Euglenophyta	1 anterior, pantonematic (more rarely more than 1 flagellum)	gullet present
Phaeophyta	2 unequal, lateral 1 acronematic, 1 pantonematic	
Rhodophyta	none	
Prasinophyceae	1, 2 or 4 flagella with fine hairs and scales on surface	cell surface covered by minute scales

* Details are given when the relevant division is considered

(b) Nuclear division does not take place by mitosis as in eucaryotic cells.

(c) The presence of a cell wall which contains a specific mucopeptide as its strengthening component (p. 27).

Thus, the less highly differentiated cells of blue-green algae can be readily distinguished from the cells of all other groups.

In several divisions the superficial appearance of the cells is distinctive and is important for the identification of such algae.

For example, in the diatoms (Bacillariophyta) the cell consists of two halves (valves), and the valve surfaces possess elaborate series of markings. Species of the Pyrrophyta also have cells of a distinctive appearance. In the main class of this division (Dinophyceae) the cell surface has characteristic longitudinal and transverse furrows with the cell surface also divided into a number of angular plates. In the other class, the Desmophyceae, the cells are divided into two halves by a longitudinal constriction.

Electron microscopic examination of the cell surface, and particularly of the surface scales, is also important for classifying Chrysophyta and Prasinophyceae.

Classification into ten phyla, each of equal status, implies that no two phyla are more closely related to each other than to any other one. This conflicts with most modern schemes of classification[26, 187, 209] which attempt to reflect supposed phylogenetic relationships. A discussion of these alternatives and the reasons for not adopting them in this book are therefore intimately connected with a general discussion of algal phylogeny, and this will be presented in Chapter 14.

2

RANGE OF VEGETATIVE STRUCTURE AND
METHODS OF REPRODUCTION

In the major part of this book each division of algae is discussed separately. Students may want to consult one particular chapter without reading the book in its entirety and in so doing they will be faced with technical terms which might be unknown to them. This present chapter aims to outline the structure and reproduction of the algae as a whole so that the meanings of some of the relevant terms can be explained.

CELL STRUCTURE

The important structural features of the cells of various algae are described separately, and no general discussion of cell structure is presented in this present chapter. However, chloroplasts can be considered, since a common, but little explained terminology is used. In most texts these structures are referred to as *chloroplasts* in some algal divisions (notably the Chlorophyta), and as *chromatophores* in all others. This distinction is generally based on differences of pigmentation; the term chloroplast being used in those species possessing chlorophylls *a* and *b* as in higher plants, and the term chromatophore being applied to species not having chlorophyll *b* and generally having an excess of carotenoids over chlorophylls. This terminology is not adopted in this book, since the structure of the organelle is basically the same in all algae (other than the blue-green algae) and is markedly different from the 'chromatophores' of photosynthetic bacteria. Thus the term chloroplast is used for all algae. The position of chloroplasts in the cell is important. They are termed *parietal* when located

towards the periphery of the cell and *axile* when located towards
the centre. A further feature of the chloroplasts which is emphasised is the presence (or absence) of a more deeply staining area
of the chloroplast generally associated with deposits of a reserve
product; this is the *pyrenoid*. The cells of archegoniate plants
generally have numerous discoid chloroplasts, and the possession
of such a feature by some algal cells is therefore emphasised.

VEGETATIVE STRUCTURE

Unicellular forms are common amongst all groups of algae except
Rhodophyta and Phaeophyta, although even amongst these
groups unicellular stages are produced at various points in their
life-histories. The unicellular species may be *motile* (*flagellate*),
non-motile (*coccoid*), or *amoeboid*.

Multicellular thalli may be regarded as being of five main types:
1. *Colonial* (coenobial) forms possessing no vegetative cell
division; 2. *Aggregations*, in which cells are aggregated into more
or less irregular colonial-like masses showing vegetative cell
division and not showing the regular differentiated construction
of the coenobial forms; 3. *Filamentous* forms also showing
vegetative division but with cells arranged in rows; 4. *Siphoneous*
form consisting of a more or less elaborate multinucleate thallus
not normally possessing septa; 5. *Parenchymatous*.

1. *Colonial*

In this type of thallus the cells are either embedded in a mucilaginous matrix, or united by a more localised production of
mucilage. It is not merely an irregular aggregation of cells but is
a well-defined colony with important reproducible features. The
coenobium is of constant size and shape for any given species, and
the cells show no vegetative division. Thus, the number of cells
of a coenobium is determined at its formation and does not
increase during growth of the colony.

2. *Aggregation*

Unlike the coenobium, an aggregation of cells is not of constant
size and shape; moreover, vegetative cell division takes place so
that there is an increase in cell number during growth. The most
common type of aggregation is the *palmelloid* form in which the
cells are embedded in an irregular mass of mucilage. The *dendroid*
colony consists of cells which are united by localised production
of mucilage to form a tree-like structure. Another kind is the

rhizopodial form of aggregation consisting of a variable number of amoeboid cells joined by a number of cytoplasmic processes.

3. *Filamentous*

Filamentous forms are also characterised by vegetative cell division but unlike the irregular aggregations the cells are arranged in linear rows. These threads, or *filaments*, are either *simple* (unbranched) or *branched*. Amongst the branched forms, a *heterotrichous* construction is one which shows a differentiation into a *prostrate* portion and an *erect* system. In many forms the heterotrichous construction is obscured by reduction or elimination of one or other of the systems.

One further modification of the filamentous habit is the *pseudoparenchymatous* construction produced by aggregation of a number of filaments to form an elaborate plant body; such a structure is the basis of all larger members of the Rhodophyta.

4. *Siphoneous*

In this, the thallus is multinucleate, but is not divided into cells (apart from those associated with reproduction). The thallus can be extremely elaborate and it is generally considered more desirable to refer to such a thallus as *acellular* and not *unicellular*.

5. *Parenchymatous*

Vegetative cell division in filamentous forms occurs in one plane so that a single row of cells is formed. When cells divide in more than one plane a parenchymatous construction is produced.

Growth of filamentous and parenchymatous thalli can be *diffuse* (all cells capable of division), *intercalary* (well-defined dividing regions not located terminally), *trichothallic* (a specialised intercalary meristem at the base of a terminal hair) or *apical* (one or more well-defined apical cells dividing to give remainder of the thallus).

METHODS OF REPRODUCTION

(The specialised processes of the Rhodophyta are omitted here.)

1. *Vegetative reproduction*

Many filamentous forms reproduce vegetatively by the fragmentation of the filament to liberate small pieces. Amongst filamentous members of the Cyanophyta this is a specialised process and short, motile lengths of filament (*hormogonia*) are formed.

2. *Asexual reproduction*

This is normally achieved by the formation of spores of various kinds. Most groups (except Cyanophyta and Rhodophyta) produce *zoospores* which are motile unicells. Non-motile asexual spores are *aplanospores*. When they appear identical to the parent cell they are referred to as *autospores*, and if they acquire a thick wall are referred to as *hypnospores*. The word 'swarmer' is commonly used for any motile cell formed when a vegetative cell reproduces, and it indicates that it is unknown (or irrelevant) whether the swarmer behaves as a gamete or as a zoospore. Amongst multicellular forms the spores may be formed in all vegetative cells, or their formation may be restricted to well-defined *sporangia*.

In the Phaeophyta two specialised kinds of sporangia can be recognised; these are the *plurilocular* and *unilocular* sporangia. The former consists of an enlarged vegetative cell which divides into a number of compartments, within each of which the contents develop into swarmers. In the unilocular type contents of the enlarged vegetative cell divide to form a number of swarmers without previous divisions of the parent cell into a number of compartments. Swarmers from the unilocular type are always asexual whereas either gametes or zoospores can be liberated from plurilocular sporangia.

3. *Sexual reproduction*

This is achieved by one of three basic methods: *isogamy, anisogamy* and *oogamy*. Isogamy involves the fusion of two morphologically identical gametes and anisogamy is fusion between morphologically dissimilar gametes. Sometimes morphologically identical gametes behave differently, and so show *physiological anisogamy*. Oogamy generally differs from anisogamy in that the female gamete is not liberated prior to fertilisation but is fertilised while still within the oogonium; an exception to this is found in the Fucales (Phaeophyceae) where the eggs are liberated from the oogonium prior to fertilisation.

4. *Germination of the zygote*

The zygote formed by all the above methods of sexual reproduction has an independent existence for a variable length of time. Generally upon germination, the contents of the zygote divide to form a number of zoospores. These are liberated and after a period of swimming they germinate into the parent plant. More rarely, the zygote germinates directly into the adult plant.

LIFE-HISTORIES OF ALGAE

During growth, an alga passes through a number of distinct phases and the sequence of these is known as its life-history. A life-history has two aspects, the somatic or morphological, and the cytological. The morphological aspect involves such questions as whether, during the life-history, the vegetative stages are morphologically alike; the cytological aspect is usually concerned with the chromosome number of each particular stage. The type of life-history generally thought to be most primitive is that in which the only vegetative stage is *haploid*, and the zygote represents the only *diploid* stage. The opposite extreme is that in which the vegetative stage is *diploid* and in which the gametes represent the only *haploid* phase. Intermediate between these two extremes are those life-histories in which there is an alternation between two vegetative stages, one haploid and the other diploid. When the two stages are morphologically similar the alternation is *isomorphic*, and when morphologically different it is *heteromorphic*.

The complexities of algal life-histories are discussed in the reviews of Drew[62] and Chapman and Chapman,[20] and both reviews introduce the complex terminology frequently used. On the basis of the nuclear phases three types of algal life-histories are sometimes recognised: *haplonts*, in which only the zygote is diploid and reduction division takes place at its germination; *diplonts*, in which only the gametes are haploid and reduction division occurs at gametogenesis; and *diplohaplonts*, in which there is an alternation of diploid and haploid stages. Drew modified this to recognise *monophasic* algae, which were either haploid or diploid, and *diphasic* algae, which corresponded to the diplohaplonts above. In addition, Drew identified three types of life-history on the basis of morphology: *monomorphic*, *dimorphic* and *trimorphic*, depending on the number of morphological stages.

The aim of such nomenclatures is to encompass all types of algal life-histories. Chapman and Chapman emphasise the deficiencies of Drew's system in this respect and so suggest further elaboration. However, Dixon[51] has pointed out how confusing the position can become, since on the basis of Chapman and Chapman's system the life-history of the red alga *Polysiphonia* is described as 'dimorphic, trigenic, dibiontic diplohaplont'! In this present book no specialised terminology is used to describe life-histories, and the essential features of the life-history are normally outlined when any particular alga is being discussed.

3

CYANOPHYTA: CYANOPHYCEAE

Division **CYANOPHYTA**
Class **CYANOPHYCEAE** (Myxophyceae;
blue-green algae)

The blue-green algae have five main features by which they are characterised. 1. Their cellular architecture is *procaryotic* (see p. 27). 2. Flagella are completely absent. 3. When movement occurs, it is by a characteristic gliding motion. 4. Their photosynthetic pigments include characteristic bilo-proteins, together with unique carotenoids such as *myxoxanthin* and *myxoxanthophyll*. 5. Storage products include the proteinaceous material, *cyanophycin*.

CELL STRUCTURE
Studies with the light microscope

When students examine the cells of blue-green algae with a light microscope they are normally impressed most forcibly by the apparent lack of any internal differentiation. The cells possess a distinct wall, sometimes distinguishable as two layers, and a mucilaginous sheath outside the wall. Inside the wall the cell contents are frequently divided into two regions, the outer peripheral *chromatoplasm* containing the photosynthetic pigments, and an inner colourless *centroplasm*. Highly refractive granules of cyanophycin are normally prominent, and numerous small dark areas representing *pseudo-* or *gas-vacuoles* can be seen in some planktonic species. There are few other features which can be recognised, and organelles such as chloroplasts, mitochondria and nuclei appear to be absent.

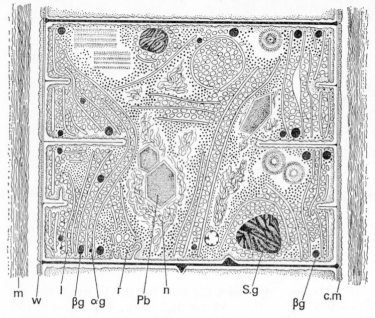

Fig. 2 Cyanophycean cell

Diagram of electron micrograph of *Symploca muscorum*. α-g, ἀ-granules; β-g, β-granules; c.m, cytoplasmic membrane; l, lamellae ('thylakoids'); m, mucilaginous sheath; n, nucleoplasm; Pb, polyhedral bodies; r, ribosomes; Sg, structural granules; w, cell wall. (After Pankratz and Bowen[185])

Studies with the electron microscope (e.g. fig. 2)

Up to the present time about 50 species of blue-green algae have been examined in the electron microscope, and, although there is some variation, a generalised picture is emerging.[21, 185, 204]

(i) *Surface layers*

The mucilaginous sheath is a constant feature of blue-green algae, although it varies in thickness from the wide one of *Anabaena*[121] to the extremely delicate one of *Anacystis montana*.[69] The sheath appears to consist of cellulose fibrils reticularly arranged within a matrix to give a homogeneous appearance.

The cell wall is inside the sheath and it consists of an outer layer (often convoluted) and an inner layer which corresponds to the 'cell wall' as seen with the light microscope. This inner layer may sometimes be further divided into two layers. The inner layer can be digested by lysozyme,[36, 86] and this, together with other evidence,[64, 85] suggests that the inner layer contains a *mucopeptide* component comparable to that found in bacterial cell walls. Essentially, a mucopeptide component consists of a peptide of a few amino acids covalently linked to two amino sugars, *glucosamine* and a derivative of this, *muramic acid*. This mucopeptide component is present in the cell walls of bacteria and blue-green algae, but is not found in the walls of any other organisms.

Inside the cell wall is the cytoplasmic membrane consisting of two electron opaque layers separated by a less opaque layer. The dimensions of this unit membrane are approximately the same as those for membrane structures of other organisms.

(ii) *Photosynthetic structures*

Under the light microscope, the chromatoplasm appears to have no internal structure, whereas examination with the electron microscope reveals a complex lamellar system,[21, 204] and these lamellae appear to be functionally analogous to the chloroplasts of other algae and higher plants. Although the lamellae of all blue-green algae investigated are of constant width, their number and arrangement varies with species, and with the age of the cell. A lamella consists of two unit membranes, each 70–80 Å thick, with a small flattened area between them. Thus they resemble the lamellae of chloroplasts of other algae and higher plants but are not separated from the remainder of the cytoplasm in a membrane-bound organelle. The lamellae probably arise by invaginations of the plasma membrane, although an alternative mechanism by which they arise in the interior of the cell has also been suggested.[21]

(iii) *Other membrane systems*

There are several independent observations[70, 185] of invaginations of the cytoplasmic membrane unconnected with the photosynthetic lamellae. These membrane systems (and the cytoplasmic membrane itself) are possibly significant because cells of blue-green algae lack mitochondria, an endoplasmic reticulum and dictyosomes (Golgi bodies). It is commonly suggested that the cytoplasmic membrane (and any invaginations thereof) is the site of biochemical functions normally associated with the well-defined membranous organelles in eucaryotic cells. However, the evidence is not conclusive, and it is to be hoped that future work with blue-green algae will clarify the position.

(iv) Nucleoplasm

The cells of blue-green algae lack an organised nucleus with a nuclear membrane and a nucleolus. Instead, the nuclear material is located towards the centre of the cell in a region of lower electron opacity than the surrounding cytoplasm. The detailed structure of the nucleoplasm depends on the methods used to prepare the material for electron microscopic examination. It is generally accepted that the image showing the fewest artifacts is that in which the nucleoplasm appears as numerous fine randomly orientated fibrils[127] (fig. 2 n). A change in the fixation technique produces an image in which the nucleoplasm appears as a transparent vacuole with a central electron opaque structure.[65] A similar change of image accompanies the same modifications of the fixative in bacteria.[98] Giesbrecht[98] interprets the dense structure in the centre of a transparent vacuole as a chromosome and suggests that the fibrillar arrangement is an artifact due to uncoiling of the chromosome. This interpretation is not, however, generally accepted since it appears unlikely that a chromosome is suspended in a completely electron transparent space; moreover, the treatment producing the fibrillar structure also preserves details of many of the cytoplasmic components.

(v) Intracytoplasmic inclusions

Although the cells of blue-green algae lack many of the organelles found in eucaryotic cells the cytoplasm is not completely devoid of inclusions. These include ribosomes, gas vacuoles (these are not morphologically equivalent to the tonoplast-bound vacuoles of eucaryotic cells), cyanophycin granules and many other granules, probably of a reserve nature but as yet uncharacterised.

(vi) Mechanism of cell division

Cleavage of the cytoplasm is initiated by a median constriction of the cell followed by introversion of the cell wall material. The cytoplasmic membrane shows centripetal growth and the inner layer of the cell wall invaginates with the membrane. The outer layer does not appear to take part in the division but is formed *de novo* between the daughter cells before they move apart. This division process is in contrast to that found in higher organisms where a new plate is formed, probably through the combined agencies of the dictyosomal material and the endoplasmic reticulum, and the new wall develops from this plate.

RANGE OF VEGETATIVE STRUCTURE

Most blue-green algae are filamentous, although unicellular and colonial forms also occur. Actual unicellular forms are rare, since the copious production of mucilage results in daughter cells remaining together after division, and irregular palmelloid forms of extreme variability are formed. Unicellular genera include *Chroococcus, Synechococcus, Anacystis* and *Gloeocapsa* (the modified classification of coccoid blue-green algae by Drouet and Daily[68] is discussed later in the chapter but the changed names of the genera are not adopted). A few cells normally aggregate together in *Gloeothece* and aggregations of numerous cells is common in such genera as *Aphanocapsa* and *Aphanothece*. Planktonic forms such as *Coelosphaerum* and *Gomphosphaera* are more elaborate colonies which consist of hollow spheres with a single peripheral layer of cells.

Amongst filamentous species it is necessary to distinguish between a *filament* and a *trichome*. A *trichome* is the basic structural unit consisting of a row of cells, whereas the filament includes the trichome plus the mucilaginous sheath. In genera such as *Oscillatoria* and *Lyngbya* each trichome is surrounded by a mucilaginous sheath (fig. 3 C) whereas in other genera (e.g. *Hydrocoleus* and *Microcoleus*) more than one trichome may be embedded in the same mucilaginous sheath, and the entire structure is termed a filament (fig. 3 A). In other genera the trichomes are embedded in more irregular masses of mucilage and no external filamentous form can be recognised. For example, in *Phormidium* the trichomes are embedded in mucilage to give an appearance of a membranous or gelatinous sheet and in *Nostoc* (fig. 3 B) the trichomes are contorted and embedded in mucilage masses which frequently form regular spheres.

In all filamentous genera mentioned above the trichomes show little differentiation; they are unbranched, there is no differentiation into a base and an apex, and growth is diffuse. Greater differentiation can be observed in members of the Rivulariaceae, in which the trichomes are whip-like with broad bases commonly attached to the substratum, the trichome tapers and usually terminates into a colourless hair. Growth is frequently *trichothallic*, that is, from a meristem of flat cells at the base of the hair.

Branching of filamentous forms is either *true* or *false*. In the former, a trichome itself branches (fig. 3 D) and is characteristic of the *Stigonematales* (e.g. *Mastigocladus*) whereas in false

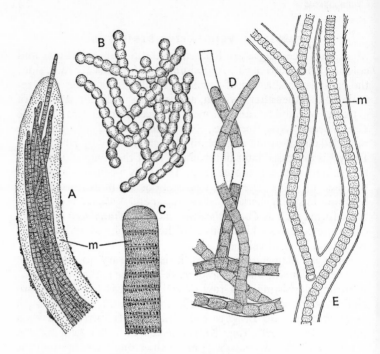

Fig. 3 Filamentous blue-green algae

A, *Microcoleus paludosus*. B, *Nostoc commune*. C, *Oscillatoria limosa*. D, *Mastigocladus laminosus*. E, *Tolypothrix fragilis*. (A–C after Fott[84]; D, E after Desikachary[42])

branching a trichome is displaced to one side and passes into a branch of the mucilaginous sheath while the other trichomes of the multiseriate filament continue in the original sheath (fig. 3 E). False branching is characteristic of the genera *Calothrix*, *Dichothrix* and *Tolypothrix*. Many of the detailed modifications of the above generalised schemes are described by Desikachary[42].

Members of the Stigonematales show the most highly developed type of vegetative construction, viz. *heterotrichy*. For example, the epiphyte *Pulvinaria* has a disc-like prostrate system and numerous erect threads. Although the two systems are sometimes visible amongst the most elaborate forms, the Scytonemataceae, most species show reduction (or loss) of either the prostrate or the erect portion.

REPRODUCTION

Vegetative reproduction by fragmentation is widespread and in the species of the Nostocales and Stigonematales specialised *hormogonia* are formed by breakage of the trichomes. Hormogonia are short lengths of trichomes with rounded ends without differentiation of the cells. In some species the breakage of the trichome appears to be related to the presence of heterocysts (see p. 32) whereas in others the trichome breaks at points where cells become modified or die.

Movement of hormogonia

Hormogonia move away from the parent plant by a characteristic *gliding* movement. Gliding has been defined as 'the active movement of an organism in contact with a solid substratum where there is neither a visible organ responsible for the movement, nor a distinct change in the shape of the organism'.[127] The mechanism of gliding is not understood and none of the theories so far advocated have been supported by definitive experimental evidence. One of the most common hypotheses relates gliding to the production of mucilage. Preliminary support for such a relationship comes from the fact that in the genus *Oscillatoria*, gliding is confined to those species which produce mucilage. Also, particles of dye move obliquely across the surface of a motile trichome in a reverse direction to the direction of gliding.[127] However, despite this indirect evidence, there is no confirmatory evidence showing a causal relationship between the localised production of mucilage and the process of gliding.

Two alternative hypotheses are of interest: 1. Using a photographic technique, Schmid[218] observed a series of waves passing along the length of the trichome suggesting rhythmic contractions. However, this has not received overwhelming support since many authorities question the interpretation of the evidence. 2. Jarosch[127] has suggested a different mechanism. This assumes there to be a fibrillar layer on the outside of the trichome and that this layer shows alternate contraction and expansion so that a wave of contraction passes along it. Jarosch suggests that the passage of such a wave would cause movement of the trichome in the opposite direction. There is little evidence for this hypothesis since it is based largely on analogy with Jarosch's observations on the mechanism of protoplasmic streaming in the cells of the Charophyceae (see p. 73). However, another school of workers[35]

speculate that the outer convoluted cell surface of the cell wall of
the Vitreosillaceae (a group of colourless filamentous gliding
organisms related to the blue-green algae) might be associated
with the gliding movement, and such a convoluted appearance
could result from the passage of a wave along the layer.

   Asexual reproduction also takes place by the production of a
number of different kinds of *spores*. 1. *Akinetes* are common in a
number of blue-green algae and are a particularly conspicuous
feature of the Nostocaceae (where they often occur in long chains)
and the Rivulariaceae. They are large cells, thick-walled and with
an accumulation of cyanophycin granules. 2. *Endospores* are produced by most species which do not produce hormogonia. During
endospore formation a vegetative cell enlarges and the contents
divide to form a number of naked endospores which show a
superficial equivalence to the aplanospores of other algae. They
are liberated and germinate immediately, without a resting stage.
3. *Nannocytes* are modified endospores, and are small naked
protoplasts formed by division of cell contents without any visible
enlargement of the vegetative cell; they are formed in such genera
as *Gloeocapsa* and *Aphanothece*. 4. *Exospores* are also modified
endospores and are produced in only two genera, *Chamaesiphon*
and *Stichosiphon*. Part of the cell wall ruptures at one point,
part of the protoplast is extruded and the exospores are pinched
off.

Heterocysts

Although heterocysts are described during this general discussion
of reproduction, the role of these structures is not understood and
it is not even known whether they are reproductive. They bear a
superficial resemblance to enlarged vegetative cells and occur
either in an intercalary or a terminal position on the trichome.
They occur in nearly all Nostocales (e.g. fig. 3 B) (except Oscilla-
toriaceae) and Stigonematales. The intercalary heterocysts usually
have a pore at each end, with protoplasmic connections between
the heterocyst and the adjacent cells; a similar pore is present at
the proximal end of a terminal heterocyst. The contents of heterocysts generally appear homogeneous and relatively dense when
young, but cyanophycin granules appear to be absent. Examination of heterocysts in the electron microscope[21, 241] reveals two
layers of the thick wall and shows that photosynthetic lamellae
are present, ribosomes are reduced in number and all other
granular inclusions are absent. It is interesting that the reduced

number of ribosomes may be related to the observation of Fogg,[80] that heterocysts become depleted of nitrogen compounds.

The function of heterocysts is unknown, although a large number of theories have been advanced; of these, three have received most widespread attention. First, it has been suggested that they function as 'points of breakage' of the trichome. This theory was suggested by the fact that breakage of the trichome into hormogonia, or during false branching frequently occurs at the heterocyst. Second, it has been thought that heterocysts stimulate the production of akinetes. This theory has developed from the frequent observation that in many species the production of akinetes is restricted to the near vicinity of the heterocysts. Fritsch[92] has developed and elaborated this hypothesis to suggest that during the vegetative stage, heterocysts secrete a substance which stimulates growth and cell division, and that during the reproductive stage the nature of the secretion changes and the production of spores is stimulated. The third hypothesis assumes that heterocysts are archaic reproductive bodies which have lost their functions. This hypothesis has developed from the observations that heterocysts sometimes germinate. However, the germination usually depends on specialised cultural conditions[245] and observations of germination under natural conditions are rare.

The constancy of heterocyst production amongst those groups which produce them argues against their being functionless, and it is this fact which prompts continued speculations and investigations of their possible function. However, they clearly remain the 'botanical enigmas' described by Fritsch.[92]

A sexual process of reproduction has not been observed in blue-green algae. However, with the identification of the parasexual process in bacteria, there is renewed interest in the possibility of finding a similar process in blue-green algae, and up to the time of writing, there are two reports[134, 220] suggesting some kind of genetic recombination in two blue-green algae.

Life-cycle of Nostoc

This section describes a single report of Lazaroff and Vishniac[138] on the development cycle of *Nostoc muscorum*. It is of possible importance, since further investigations with a similar technique might indicate that, despite simplicity of reproductive methods, blue-green algae show a well-defined cycle of development. From studies of the effect of culture conditions on the growth and reproduction of *N. muscorum*, Lazaroff and Vishniac concluded

B

that there was an alternation between a *heterocystous* cycle and a *sporogenous* cycle. The former cycle occurs under normal phototrophic conditions (i.e. illumination with carbon dioxide as carbon source) and consists of the production of hormogonia by breakage of trichomes at heterocysts and the subsequent development of these hormogonia into heterocyst-containing trichomes from which the cycle begins again. When the organism is grown in the dark with glucose or sucrose as the carbon source they break up into individual cells. At first clumps of cells are formed and later, short motile filaments (i.e. morphological equivalents of hormogonia) are produced. That is, these hormogonia are produced without the organism first passing through a heterocyst stage. When these motile filaments are illuminated they produce heterocystous filaments after a period of anastomosis.

CLASSIFICATION

It is proposed to outline the classification system of Desikachary[42] since this summarises the more established systems. There are, however, several features of the classification of blue-green algae which make one question the validity of some of the criteria used; these features are discussed later in this section.

The blue-green algae are usually divided into five main orders: *Chroococcales, Chamaesiphonales, Pleurocapsales, Nostocales,* and the *Stigonematales*.

Members of the *Chroococcales* are unicellular or colonial and most are contained in the family Chroococcaceae (e.g. genera such as *Anacystis, Synechococcus, Gloeocapsa* and *Aphanocapsa*). The order never shows a trichome organisation although a pseudo-filamentous colony is formed in species of the Entophysalidaceae (e.g. *Chlorogloea* and *Entophysalis*). Amongst all species of the order the method of reproduction is usually a vegetative process of cell division or colony fragmentation. Endospores are never produced although nannocytes are sometimes formed.

The *Chamaesiphonales* also include unicellular (e.g. *Dermocarpa*) and colonial (e.g. *Chamaesiphon*) forms, but they differ from the Chroococcales in that they produce endospores (or exospores in *Chamaesiphon*). *Stichosiphon* has a superficial resemblance to a uniseriate filament. This appearance arises from segmentation of an elongated cell during formation of a row of endospores.

The three remaining orders are filamentous. Of these, the

Pleurocapsales differ from the others in lacking hormogonia. Instead, its members reproduce by the production of endospores. Although most species are heterotrichous, there is usually a reduction of either the prostrate or the erect system.

Nostocales and *Stigonematales* are the two largest groups of blue-green algae and both are characterised by the production of hormogonia. Nostocales are unbranched (e.g. *Oscillatoria, Lyngbya*), or show false branching (*Scytonema, Plectonema*), whereas true branching is characteristic of the Stigonematales (e.g. *Mastigocladus*).

Although the general classification of blue-green algae into orders and families as outlined above is generally accepted by most authorities, the criteria used for the classification of blue-green algae at the generic and species level is in a confused state. For example, Drouet and Daily[68] have proposed a revised classification of the coccoid forms and suggest that the Chroococcaceae should include only five genera. This contrasts with the 14 genera described by Desikachary.[42] The revised scheme of Drouet and Daily[68] is based on their observations of the effect of changes in environmental conditions on the form of the Algae. Drouet[67] investigated similar changes on the form of the unbranched filamentous alga *Schizothrix calcicola*. After comparing the various forms with type specimens of other species and genera Drouet concluded that environmental changes produce forms which are indistinguishable from 54 taxa from genera such as *Amphithrix, Lyngbya, Oscillatoria, Phormidium, Plectonema, Pseudoncobyrsa, Schizothrix, Symploca* and *Tapinothrix*. He suggests that all these taxa are ecophenes (that is, ecological growth forms) of a single species, synonymous with *Schizothrix calcicola*. Similar ecophenes of *Microcoleus vaginatus*[66] have also been found and the alga appears to be equivalent to *Phormidium* spp., *Hydrocoleus homoeotrichium*, and *Lyngbya aerugineo-caerulea*. Clearly, further investigations under controlled environmental conditions are needed before the classification of blue-green algae can be completely revised.

Relationship between blue-green algae and bacteria

Like most groups of algae, the Cyanophyta contain several colourless forms (e.g. *Beggiatoa*). These bear a strong superficial resemblance to some bacteria. Because of this, Cohn[30] suggested that both bacteria (Schizomycetes) and blue-green algae (Schizophyceae) be regarded as two classes of the phylum, Schizophyta.

During the sixty to seventy years following the publication of Cohn's ideas more colourless forms of blue-green algae were observed but the significance of their superficial resemblances to bacteria was disputed. In 1941 Stanier and van Niel[224] concluded that there the two groups were related, with three main features in common; absence of nuclei, absence of sexual reproduction and absence of plastids. However, in 1949, Pringsheim chose[201] to emphasise the differences between them, and also pointed out that the characters enumerated by Stanier and van Niel were negative, and so questioned their significance. More recently, electron microscopic examination of the cells has necessitated a re-examination of the problem since the negative features of Stanier and van Niel can now be replaced by more positive characters. In addition to a procaryotic cell structure (see p. 17), they possess several biochemical features in common; for example, their method of ornithine biosynthesis and their sensitivity to certain antibiotics. The features shared by bacteria and blue-green algae are reviewed in detail by Echlin and Morris,[71] and it is clear from such details that bacteria and blue-green algae possess several distinctive features in common and which warrant their separation from all other cellular organisms. The terms of Christensen,[26] *Procaryota* and *Eucaryota*, seem excellent terms for this primary classification of cellular organisms.

4

CHLOROPHYTA: CHLOROPHYCEAE

Division CHLOROPHYTA
Class CHLOROPHYCEAE (Green algae)

The Chlorophyceae is an extremely large class of algae, and the range of vegetative structure is greater than that in any other algal class. Methods of reproduction are also diverse. Despite the extreme variability of green algae, they have four main features in common. 1. The photosynthetic pigments include chlorophylls *a* and *b* (the Euglenophyta is the only other algal division possessing chlorophyll *b*). 2. Starch accumulates as a reserve product. 3. The cell walls invariably contain cellulose. 4. Flagella, when present, are normally 2 or 4 in number and all are of the acronematic type. Most green algae are freshwater, although they are also common constituents of the soil flora and a few are marine.

Because the class is so large, a single chapter in a small book must necessarily omit many details, and attempt, instead, to provide a skeleton outlining the more important features of the group. In so doing, any particular species or genus is mentioned only to illustrate the appropriate point under discussion, and no single species of alga is described in detail.

The general approach to the construction of the skeleton is one based on the vegetative form of the Algae. That is, the discussion will be based on increasing complexity of the plant body and this will be correlated with any parallel changes in the functional specialisation, method of reproduction, and life-history. Thus, the approach adopts the basic reasoning of Blackman[11] who suggested that within the green algae one could identify 'tendencies', that is, changes of structure and reproduction reflecting evolutionary lines of development. Moreover, Blackman also observed that the *order* was the basic taxonomic unit and that most of the important 'tendencies' could be recognised by reference to appropriate orders. The status of the order assumes comparable importance in this present chapter, and although individual genera and species are mentioned in relation to particular details, taxonomic units of a rank lower than order (sub-order, family, sub-family, etc.) are not normally considered.

It must be emphasised that the 'lines of development' to be described in this chapter are designed to clarify the understanding of an extremely large group of organisms, and their possible relationship to the evolution of the group is of secondary importance.

For the purpose of this present chapter the Chlorophyceae are divided into 14 orders: Volvocales, Tetrasporales, Chlorococcales, Ulotrichales, Chaetophorales, Sphaeropleales, Ulvales, Oedogoniales, Zygnematales, Cladophorales, Acrosiphonales, Siphonocladales, Dasycladales and Caulerpales (Siphonales under the old nomenclature).

Classification of the Chlorophyceae is undergoing rapid changes and the discussion by Round[208] is recommended for detailed modifications of the present scheme. Although new classification schemes may be widely accepted in the future, at present they depend on 'weighting' one or two features and using these to create new classes and subclasses. Thus, although the new schemes are exciting, and promote further work, they are not yet widely accepted and so are not presented here.

The 14 orders can be conveniently studied by considering them in three main groups: 1. the unicellular and colonial forms, 2. the uninucleate filamentous forms, and 3. the multinucleate or siphoneous forms.

UNICELLULAR AND COLONIAL FORMS
(VOLVOCALES, TETRASPORALES, CHLOROCOCCALES)

The basis of separation of these 3 orders is as follows: the vegetative stages of the Volvocales are motile, whereas those of the Chlorococcales and the Tetrasporales are not. Chlorococcales contain both unicellular and colonial forms whereas all Tetrasporales are multicellular. The multicellular forms of the two orders can be distinguished by the fact that those of the Chlorococcales do not exhibit vegetative cell division, whereas those of the Tetrasporales do.

Volvocales

The unicellular species of this order are very variable. In most species of *Chlamydomonas*, e.g. *C. angulosa*, the cell is bounded by a firm cell wall and two acronematic flagella project through the wall at the anterior end. Most of the cell is occupied by a single basin-shaped chloroplast with a single pyrenoid towards the posterior, and an eye-spot (stigma) is located towards the anterior of the cell (fig. 4 A). It is not possible to present all variations of this picture, but the main ones can be recognised by reference to four features:

(i) *Cell wall*

A cell wall is absent from some genera, for example *Polytomella*. In others a *lorica* is present (e.g. *Phacotus*, *Dysmorphococcus*). In these, the cell surface is separated from the outer firm envelope by a relatively large space (fig. 4 B). In two other genera, *Haematococcus* (Sphaerella) and the colonial *Stephanosphaera* the cell wall is separated from the protoplast by a mucilaginous space, transversed by strands of cytoplasm (fig. 4 C).

(ii) *Flagella*

Although most genera are biflagellate, *Carteria* (for example) is quadriflagellate. In all genera the flagella are acronematic, of equal length and located at the anterior of the cell.

(iii) *Chloroplasts*

Although the basin-shaped chloroplast is most common, reticulate peripheral chloroplasts are present in *Haematococcus lacustris* and *Chlamydomonas reticulata*; axile stellate chloroplasts are found in *Chlamydomonas arachne* and *C. eradians*, and small discoid chloroplasts in *C. alpina*. Colourless genera such as *Polytoma*, *Polytomella*, *Hyaliella* are known.

(iv) *Pyrenoids*

Although a single pyrenoid is most commonly present, there is some variability. *Chloromonas* does not have a pyrenoid, whereas *Chlamydomonas sphagnicola* has numerous pyrenoids irregularly distributed throughout the chloroplast.

The above remarks have emphasised the gross structure of the cell and the main types of variation recognisable with the light

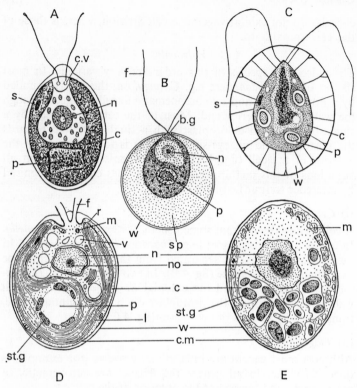

Fig. 4 Unicellular Volvocales

A, *Chlamydomonas angulosa*. B, *Phacotus lenticularis*. C, *Haematococcus pluvialis*. D, *Chlamydomonas* (species unspecified), diagram of electron micrograph. E, *Polytoma uvella*, diagram of electron micrograph. b.g, basal granules; c, chloroplast; c.m, cytoplasmic membrane; c.v, contractile vacuoles; f, flagella; l, lamellae; m, mitochondria; n, nucleus; no, nucleolus; p, pyrenoid; s, eye-spot; sp, space between wall and cytoplasm; st.g, starch grain; v, vacuole; w, wall. (A after Fritsch[90]; B, C after Fott[84]; D after Sager[211]; E after Lang[137])

microscope. Before leaving the structural aspects of unicellular species of the Volvocales, several interesting features which have emerged from more recent investigations should be mentioned.

(i) *Electron microscopic examination of a Chlamydomonas cell*[213] (fig. 4 D)

The cell is bounded by the cytoplasmic membrane which appears to be a typical unit membrane of two electron opaque layers separated by a less opaque zone. Outside this is the cell wall which is thickened and stratified at the anterior pole. Each flagellum shows the typical 9+2 pattern of component fibrils similar to that in all plant and animal flagella. Each flagellum is attached inside the cell to a basal granule. The large chloroplast is surrounded by a unit membrane and within it the following structures can be identified: (a) *Lamellae* occur in pairs, each pair forming a closed system (disc) and the discs occur in stacks of 2–20. (In other algal divisions the chloroplast is said to be traversed by 'bands' and this term is equivalent to 'stack'. That is, there is only imperfect segmentation into grana). This structure of the chloroplast is general for almost all algae, although the number of discs per band (stack) is variable.[95] The number of discs constituting a band is believed to be of phylogenetic significance since there is only one disc in the bands of red algae (see p. 149) and in *Cyanidium* (see p. 133), whereas in all other algal divisions there is more than one disc in a band, and the largest numbers are found in the Chlorophyceae. (b) A *granular matrix*. (c) *Starch grains*. (d) A *pyrenoid*, with a granular core surrounded by tightly packed starch plates. The core of the pyrenoid is not traversed by the lamellae (this observation is similar to that in *Chlorella*, *Spirogyra* and *Closterium* (all green algae) but is different from that in *Euglena*). (e) An *eye-spot* consisting of 2 (sometimes 3) curved plates concentrically disposed near the anterior of the chloroplast. Each plate shows hexagonal packing of uniform bodies (cf. *Euglena*, p. 123).

Other cytoplasmic structures include the nucleus, mitochondria, endoplasmic reticulum and dictyosomes (Golgi bodies).

(ii) *Function of the stigma and contractile vacuoles*

The stigma is generally accepted to be the light receptor for the phototactic response of the cell. However, a stigma-less mutant strain of *Chlamydomonas* responds to light, but does so less quickly and less uniformly.[110] Thus, although it is important it is not the sole means of light perception in this alga.

The contractile vacuole is generally assumed to be responsible for the removal of water from the cell. Evidence for such a function in *Chlamydomonas* comes from studies with an X-ray-induced mutant which lacks a contractile vacuole.[106] Such a mutant could only survive in solutions with an osmotic pressure of 1·5 atmospheres. Moreover, when the high osmotic pressure is achieved by compounds such as ethylene glycol (a substance which enters the cell readily) the cells did not survive. When the wild-type cells are grown in medium with an osmotic pressure of 1·5, contractile vacuoles do not form.

(iii) *Electron microscopic examination of* Polytoma

Polytoma is an example of a colourless alga, and the basic question raised by such algae is whether the loss of photosynthetic pigments is accompanied by the loss of the chloroplast. Lang[137] has observed a membrane-bound area in the posterior part of the cell of *Polytoma* apparently representing a single cup-shaped chloroplast (fig. 4 E); but no lamellar structures can be observed. This presence of plastids in colourless cells contrasts with that found in Euglena (see p. 124).

(iv) *Mechanism of flagellar movement*

As yet there is no comprehensive scheme for the mechanism of flagellar movement in algae, but more work has been done with *Chlamydomonas* and *Polytoma* than with any other algae. This work has been reviewed by Brokaw[18] and the main observations on the mechanism of flagellar action are as follows:

(a) The two flagella are probably curved towards the back of the cell immediately after emerging from the anterior end. No analysis of the pattern of flagellar movement has yet been made, but from analogy with sea-urchin sperm it is suggested that a sine wave passing from the base to the tip of the flagella is responsible for the forward movement of the cell (note that this conflicts with earlier ideas of a 'breast-stroke' type of movement).

(b) Isolated flagella can beat, if supplied with adenosine triphosphate (ATP), they exhibit ATPase activity and such activity is absent from paralysed mutants.

(c) The amino acid composition of the flagellar protein shows many similarities with that of actin or myosin from muscle cells.

From these studies one can speculate that biological movement involves contraction and expansion (cf. the gliding hypothesis of Jarosch, p. 31) carried out at the expense of ATP, and that in flagellar movement the alternate expansion and contraction produces a sine wave passing from the base to the tip of the flagellum. However, many observations with other algal groups do not agree with such a comprehensive view of flagellar movement.

Reproduction of unicellular forms

Most commonly, *asexual reproduction* is achieved by division of the cell contents and liberation of a number of daughter cells. The parent cells normally lose their flagella before the onset of division.

Sexual reproduction is common in unicellular members of the Volvocales, and is usually achieved by division of a vegetative cell to form a number of gametes which are liberated as naked free-swimming cells. Isogamy is common, although morphological isogamy is frequently accompanied by physiological anisogamy in which the behaviour of the gametes might differ. Also some species of *Chlamydomonas* are heterothallic, and fusion only occurs between gametes from different strains. Morphological anisogamy has been confirmed in *Chlamydomonas braunii* where 4 larger macrogametes are liberated from some cells and 8 small microgametes from others. Simple oogamy occurs in some species. For example, in *Chlorogonium oogamium*, a single naked amoeboid protoplast is set free as an ovum and a large number of small spermatozoids are liberated by another cell. In *Chlamydomonas coccifera* a vegetative cell loses its flagella, rounds off and enlarges; this ovum is fertilised by male gametes formed (usually 16) in other cells.

The details of the fusion process have been examined in many species of *Chlamydomonas*. Normally, when gametes from compatible strains are mixed they are attracted to each other, clumps of cells are formed, and these later break up so that the gametes swim off in pairs. The contact of the cells in pairs is achieved by agglutination of the flagella (they appear to stick and not to intertwine). A second surface reaction is responsible for actual pairing of the cells by contact between the cell surfaces. Wiese and Jones[240] have shown that the mating reaction involves only the flagellar surface reaction since one can observe the reaction with isolated flagella.

The membrane breaks down between the two cells, cytoplasmic fusion is followed by nuclear fusion and a cell wall is formed. The wall of the zygote thickens and it usually remains dormant for some time before germinating. Meiosis accompanies zygote germination and 4 haploid cells are liberated.

Colonial Volvocales

The motile colonies have a fixed number of cells arranged in a constant and reproducible way. These colonies are termed

coenobia. The important feature about the coenobial construction is that the number of cells in the colony is determined at the
beginning of their development, and during growth the cell
number does not increase.

Colonial Volvocales show a striking series of increasing
elaboration and specialisation, and this series can be described
by reference to five genera: *Gonium, Pandorina, Eudorina, Pleodorina,* and *Volvox* (figs. 5 A–E). There are four main features
showing the increased elaboration: coenobial size and shape,
morphological differentiation of the cells, specialisation of reproductive cells and the method of sexual reproduction.

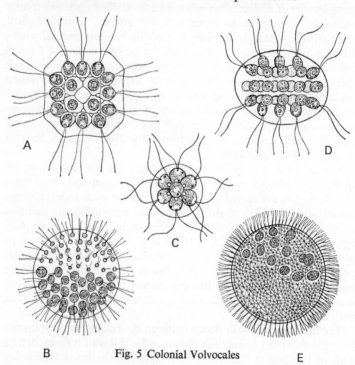

Fig. 5 Colonial Volvocales

A, *Gonium pectorale.* B, *Pleodorina californica.* C, *Pandorina morum.*
D, *Eudorina elegans.* E, *Volvox aureus.* (After Fott[84])

Coenobial size and shape

Gonium has 4, 16 or 32 cells (the actual number depends on
species) arranged as a flat, or curved, plate; *Pandorina* colonies

also have 4–32 cells but the colonies are spherical, as are the colonies of *Eudorina*, *Pleodorina* and *Volvox*. *Eudorina* has 16, 32 or 64 cells, *Pleodorina* has 32–128 cells and *Volvox* species vary from 500–50,000 cells.

Structural differentiation

All vegetative cells of the *Gonium* colony appear identical, and the colony moves by a somersaulting motion, so that the colony has no distinct anterior and posterior ends. All other genera being described have distinct anterior and posterior ends, and this can be recognised not only by the motility, but by structural differences between the anterior and posterior cells. For example, in *Pandorina* and *Eudorina* the stigma tends to become less prominent in the posterior cells, and it is absent from the posterior cells of *Pleodorina* and *Volvox*. In both *Volvox* and *Pleodorina* the posterior cells are larger than the anterior.

Asexual reproduction

Asexual reproduction is effected by division of some of the vegetative cells into a number of daughter cells equal to the number of cells in the coenobium. In *Gonium*, *Pandorina* and *Eudorina* all the vegetative cells function in the asexual process, whereas in *Pleodorina* only those cells in the posterior half function, and in *Volvox* only a few specialised cells, *gonidia*, divide (fig. 5 E). The details of division and subsequent arrangement of the daughter cells as a new colony are similar in all genera, and can be found in most of the standard texts. The important fact is that after division the daughter protoplasts become arranged in a curved plate, the *plakea*, before being arranged in the form of the parent colony.

Sexual reproduction

The sexual process of *Gonium* is relatively simple. The cells of the colony divide as in asexual reproduction but in the sexual colony the naked daughter protoplasts function as gametes, and fusion is isogamous. Although the gametes are morphologically identical they presumably show some physiological differences since some species of *Gonium* are heterothallic.[228] *Pandorina* is also heterothallic. The formation and fusion of the gametes resemble that in *Gonium*, except that gametes from different strains frequently differ in size. Only some of the cells of *Eudorina* divide to produce gametes, and the fusion is either anisogamous or oogamous. The spindle-shaped antherozoids are liberated in a single

unit, they swim as a unit before breaking up into separate antherozoids which fertilise vegetative cells functioning directly as
female gametes. In *Pleodorina* the method is similar except that
only a few larger cells function, and sometimes the flagella of the
female gamete are lost. In *Volvox* the unit of antherozoids
(16–512) is liberated as in *Eudorina* and the female gamete is an
enlarged cell which has lost its flagella.

Germination of the zygote is accompanied by reduction
division of the nucleus and in all the above genera, except *Gonium*,
3 of the 4 daughter nuclei degenerate and a single zoospore is
liberated to form the adult colony. In *Gonium*, however, the 4
nuclei remain. a 4-celled colony is liberated, and subsequently
each of the 4 cells divides to form a daughter colony.

As in the sexual process of unicellular forms the first step in
gamete fusion is the agglutination of flagella, and Coleman[33, 34]
has used the various mating types of *Pandorina morum* to investigate the possibility that this agglutination is comparable to
an antibody-antigen reaction. After preparing antisera to
flagellar preparations from different mating types, it was found
that an antiserum caused shedding of the flagella in the immunising strain and in its complement, but significantly, had no
effect on strains incompatible with the immunising strain.

Chlorococcales

In this order the normal vegetative cell is non-motile. As in the
Volvocales one can recognise unicellular and colonial forms and
our discussion of these will follow the approach used for the
Volvocales. However, one additional important development is
the tendency for the cells to become multinucleate.

Structure of unicellular forms

Most species of *Chlorococcum* consist of spherical cells with a
well-defined cell wall, a single nucleus and a large parietal chloroplast occupying a large area of the cell periphery and containing
a pyrenoid (fig. 6 A). Thus the structure resembles most species
of *Chlamydomonas* except that flagella, eye-spots and contractile
vacuoles are absent. Although the absence of these organelles is
general for non-motile cells, contractile vacuoles are present in
the non-motile *Hypnomonas*.

Although most species of *Chlorococcum* have parietal chloroplasts, *Trebouxia* has a large lobed *axile* chloroplast with a
pyrenoid, *Dictyococcus* has a number of plates and no pyrenoids,

and *Eremosphaera viridis* has a large number of small chloroplasts also devoid of pyrenoids.

Asexual reproduction of the unicellular Chlorococcales is achieved by division of the cell contents to produce 8 or 16 naked,

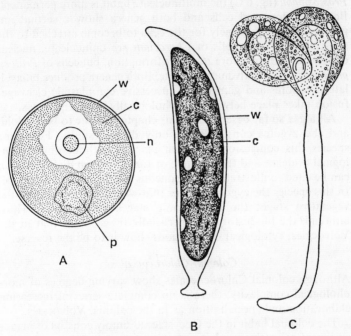

Fig. 6 Unicellular Chlorococcales

A, *Chlorococcum multinucleatum*. B, *Characium apiculatum*. C, *Protosiphon botryoides*. c, chloroplast; n, nucleus; p, pyrenoid; w, wall.
(A, C after Fott[84]; B after Fritsch[90])

biflagellate, ellipsoidal zoospores. At the onset of zoospore formation the nucleus divides until 8–16 daughter nuclei are produced, and successive cytoplasmic cleavage produces the zoospores. Thus, for a short time, the cells are multinucleate. This is general for the Chlorococcales and contrasts with the Volvocales, where each nuclear division is followed immediately by cytoplasmic cleavage. Some species are azoosporic, that is, they reproduce by the production of aplanospores instead of zoospores. The genus *Chlorella*, an alga used widely for physiological investigations, is an example of such azoosporic forms.

Chlorococcum reproduces sexually by an isogamous process, whereas no sexual process has yet been identified in *Chlorella*.

The tendency for cells to become multinucleate for a short time has already been observed, but in *Characium* (fig. 6 B) and *Prostosiphon* (fig. 6 C) the multinucleate habit is more permanent. Both are elongated cells and both genera show a second important tendency, namely for the cells to become attached to the substratum. Young cells of *Characium* are uninucleate, nuclear division occurring before zoospore formation, but cells of *Protosiphon* are permanently multinucleate. Both genera produce biflagellate zoospores and gametes by successive cytoplasmic cleavage; fusion takes place between morphologically similar gametes.

All algae so far described in this chapter appear to be haploid, and the zygotes constitute the only diploid stages. For some species this conclusion has been reached without direct cytological evidence and the example of *Chlorococcum diplobionticum* can be cited to illustrate how erroneous such conclusions may be. In this species the zygote remains thin-walled and constitutes the vegetative stage; the only motile stages are gametes. Thus, although the life-history is superficially the same as that in the Volvocales, cytological observations show it to be the reverse.

Colonial Chlorococcales

Although colonial Chlorococcales show varying degrees of morphological complexity, there is no complete series of increasing elaboration and specialisation as in the colonial Volvocales.

The colonial habit in the Oocystaceae simply consists of aggregation of cells (the number and shape is indefinite) in a mucilaginous envelope. Members of this group reproduce asexually by the production of *autospores* and the colonial habit results from failure of these spores to separate. Thus the boundary of the mucilaginous colony is the wall of the original parent cell. Another group, containing such genera as *Golenkinia* and *Micractinium*, also has colonies formed by failure of autospores to separate. However, in these genera the colonies are generally more fixed in size and shape. Moreover, in addition to reproducing asexually by autospore production, they have an oogamous sexual process, during which the contents of some cells divide into 8–16 biflagellate antherozoids, whereas in the female colony a vegetative cell functions as an ovum and is fertilised through a pore which develops when the ovum ripens.

In other colonial forms, the colonies are more regular and are well-defined coenobia. The main features of these forms can be illustrated by reference to three common genera: *Scenedesmus*, *Pediastrum* and *Hydrodictyon*. *Scenedesmus* is a genus widely used in physiological investigations and consists of a linear series of 4 or 8 cells (changes in nutritional conditions can alter the number from 2 to 32) joined together by the localised production of mucilage (fig. 7 B), and the outer cells of the series frequently have spines or processes. During asexual reproduction a cell divides into a number of autospores which become orientated in the form characteristic of the particular species. *Scenedesmus* is generally quoted as an example of azoosporic Chlorococcales, reproducing

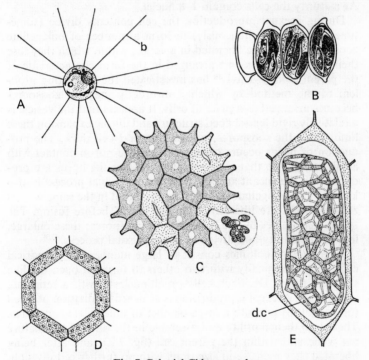

Fig. 7 Colonial Chlorococcales

A, *Micractinium radiatum*. B, *Scenedesmus quadricauda*, reproducing coenobium. C, *Pediastrum boryanum*. D, *Hydrodictyon reticulatum*, detail of part of net. E, *H. reticulatum*, a new net formed within an enlarged cell. b, hollow bristle; d.c, daughter colony. (A, B after Fritsch[90]; C, D, E after Fott[84])

by the production of autospores and never producing motile stages. Recently, Trainor[236, 237] has observed biflagellate motile stages when cultures of *Scenedesmus obliquus* are maintained under conditions of nitrogen deficiency. Moreover during culturing, the *Scenedesmus* appeared to pass through stages resembling other Chlorococcacean genera, *Dactylococcus*, *Chlorella*, *Oocystis* and *Ankistrodesmus*. This affords another example of the way in which studies of algae in culture frequently invalidate many of the criteria used to identify given genera.

Pediastrum is a common genus in some plankton samples and consists of a flat plate of cells (2–128, depending on species) and the outer ones normally have projections on the surface (fig. 7 C). At maturity the cells contain 1–8 nuclei.

During asexual reproduction the cell contents divide (sometimes all cells simultaneously) to form a number of biflagellate zoospores. These are liberated in a vesicle within which they lose their flagella and become arranged in the form characteristic of the parent colony. Davis[39] has investigated the interesting problem of the method by which a disorderly mass of zoospores becomes arranged as a plate of cells. It appears that the vesicle is a relatively rigid lens-shaped structure and during expansion these limits force the zoospores to form a layer 1 cell thick. The production of horns occurs on any cell surface not in contact with adjacent cells, so that in interrupted colonies the horns are produced into the vacant areas. An isogamous sexual process is also known. The biflagellate gametes are produced in the same way as zoospores but are liberated from the vesicle before fusing. The zygote germinates by the formation of zoospores; these enlarge, lose their motility and divide as in the asexual process.

Hydrodictyon colonies consist of large numbers of cylindrical cells joined terminally with two others to form an open net-like structure (fig. 7 D). Each cell is multinucleate with a reticulate chloroplast. Asexual reproduction is achieved by division of a cell (or several) to produce a large number of biflagellate zoospores. These lose their motility and assemble in the form of the mature net while still within the parent cell (fig. 7 E); and after being liberated they increase in size only, with no further cell division. Cells appear to grow where in contact with others and do not grow when not touching adjacent cells.[170] Isogamous sexual reproduction is also known. It appears that the nuclear division at the onset of zygote germination is not meiotic, and the vegetative plant is therefore assumed to be diploid.

Tetrasporales

When *Chlamydomonas* is grown on agar the cells are unable to swim away from each other and so remain as irregular gelatinous masses. In some algae these *palmelloid* stages constitute the normal vegetative phase and these forms are included in the *Tetrasporales*. Their separation from the Chlorococcales is based mainly on the presence or absence of vegetative cell division. Vegetative division is that which is not associated with reproduction and generally involves division of a cell into two daughter cells. This contrasts with the division of a *Chlamydomonas* or *Chlorococcum* (for example) cell into a number of spores which are then liberated. On the basis of this difference, Fritsch[90] defines vegetative cell division as that in which there is no rupture or gelatinisation of the parent cell wall. Species of *Chlorococcales* do not normally show such vegetative division, whereas it can occur in species of the Tetrasporales. Herndon[117] disagrees with this interpretation, and considers that there is no difference between cell division in the Tetrasporales and the Chlorococcales, and concludes that almost all non-motile species should be included in the same order, Chlorococcales; and the few exceptions showing true vegetative cell division should be separated in a distinct order, *Chlorosphaerales*.

This practice is not adopted here since the distinction between Chlorococcales and Tetrasporales leads to few ambiguities. Also, in addition to vegetative cell division, other differences between the two orders are common (but not absolute). For example, the vegetative cells of Tetrasporales sometimes acquire flagella and become motile; this never happens in the Chlorococcales. Also, cells of the Chlorococcales are sometimes multinucleate, at least for part of their life-history, whereas the cells of the Tetrasporales are never multinucleate.

Palmella is a genus of the Tetrasporales consisting of an unspecialised aggregation of cells in a gelatinous matrix. The cells sometimes acquire flagella and swim away. Asexual reproduction is achieved by division of the contents of some cells into 4 zoospores; sexual reproduction occurs by isogamous fusion of gametes produced in the same way as zoospores. The planktonic *Sphaerocystis* is a related genus in which the aggregations are generally more fixed in size and shape. *Tetraspora* is slightly different; it occurs as elongated slimy masses within which the cells are embedded. A long cytoplasmic process passes from the face of each cell to the surface of the cell and there projects as a *pseudocilium*.

FILAMENTOUS AND PARENCHYMATOUS FORMS

The filamentous habit is characterised by vegetative cell division which normally occurs in only one plane. There are 7 orders showing such a construction: *Ulotrichales, Chaetophorales, Oedogoniales, Zygnematales, Sphaeropleales, Acrosiphonales,* and *Cladophorales.* The cells of the last three are multinucleate. However, there is evidence[214] to suggest that the Sphaeropleales are related to the Ulotrichales; whereas the Cladophorales and Acrosiphonales are best discussed later in the chapter with the three multinucleate orders, Dasycladales, Siphonocladales and the Caulerpales.

There are also two orders containing parenchymatous forms, the Ulvales and Prasiolales. These are relatively simple, however, and can be conveniently discussed with the filamentous forms.

Ulotrichales, Sphaeropleales, Chaetophorales, Ulvales, Prasiolales

Within these 5 orders it is possible to identify a number of important 'tendencies' or 'lines of development' and these can best be illustrated by beginning with the genus *Ulothrix* (Ulotrichales). For the purpose of the present discussion the essential features of most species of *Ulothrix* are as follows: the filament is *simple* (unbranched) with little differentiation of the cell structure, except that the basal cell is usually colourless and modified as an attaching cell; all other cells have a characteristic girdle-shaped chloroplast (fig. 8 A). Growth is diffuse (any cell is capable of division) and, apart from fragmentation, *Ulothrix* reproduces asexually and sexually by the production of a variety of swarmers. Thus *U. zonata* produces 3 kinds of swarmers: 1. quadriflagellate macrozoospores (more common), 2. quadriflagellate and biflagellate microzoospores, and 3. biflagellate gametes. Aplanospores and akinetes are also produced under suitable conditions. The various swarmers are produced by normal vegetative cells and sexual fusion is isogamous. The life-history is simple, consisting of a haploid plant body with the zygote as the only diploid stage. The developments which the other filamentous forms are likely to show can be summarised as follows:

1. *Within the Ulotrichales:*   (a) modifications of cell structure
                                      (b) development of an oogamous
                                          sexual process

2. *Cells become multinucleate* (Sphaeropleales)
3. *Within the Sphaeropleales:* (a) modifications of cell structure
(b) development of oogamy
4. *The development of branched filaments* (Chaetophorales)
5. *Within the Chaetophorales:* (a) all are heterotrichous
(b) structural modifications
(c) development of specialised sporangia
(d) development of oogamy
6. *Development of the parenchymatous construction* (Ulvales, Prasiolales)
7. *Within the Ulvales:* development of an isomorphic alteration of generations
8. *Within the Prasiolales:* (a) different cell structure
(b) specialised life-history

The above summary of developments is an oversimplification in the sense that there are numerous other combinations between the various alternatives of cell structure, vegetative forms, methods of reproduction and life-history. However, it serves as a useful skeleton on which to base the more detailed discussion.

Cell structure in the Ulotrichales

There is much variation in the form of the chloroplast. For example, *Hormidium* is distinguished from *Ulothrix* by the fact that the chloroplast is much smaller, occupying only half the length of the cell. The cell wall shows two important variations. In some species of *Microspora* the wall is composed of two halves, the ends of which overlap each other, and characteristic 'H' pieces can be identified (fig. 8 B). In *Cylindrocapsa*, each elliptical cell has a thick stratified wall, and in some of the older filaments oblique divisions may occur so that parts of the filament are not uniseriate (fig. 8 C).

Oogamous reproduction in the Ulotrichales

Cylindrocapsa is the one genus of this order with a well-defined oogamous process. Antheridia are formed by rapid successive division of certain cells to give a series of small cells, each of which produces 2 elongated biflagellate antherozoids. The cell developing into the oogonium enlarges and acquires a thick stratified wall. Within this structure the single ovum is formed by contraction of the protoplast and is fertilised through a pore in the lateral wall (fig. 8 C).

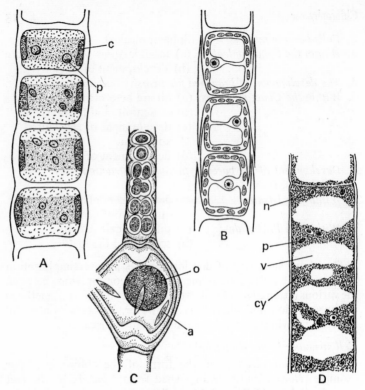

Fig. 8 Ulotrichales and Sphaeropleales

A, *Ulothrix zonata*. B, *Microspora quadrata*. C, *Cylindrocapsa involuta*. D, *Sphaeroplea annulina*. c, chloroplast (girdle-shaped in *Ulothrix* and reticulate in *Microspora*); cy, lining cytoplasm with annular chloroplasts; n, nuclei; p, pyrenoid; v, vacuole; a, anthero-zoid; o, oogonium. (A-C after Fott[84]; D after Fritsch[90])

Cell structure of the Sphaeropleales

The cells of *Sphaeroplea* are much longer than broad (contrast *Ulothrix*) and contain several transverse cytoplasmic septa separated from each other by large vacuoles (fig. 8 D). Several nuclei and a band-shaped chloroplast (or a number of discoid chloroplasts) are present in each cytoplasmic septum, but chloro-plasts do not appear to occur in the narrow lining layer of cytoplasm around each vacuole.

Recent cytological observations by Sarma[214] have shown that the nuclear structure of *Sphaeroplea* is similar to that of the Ulotrichales but is very different from that of the Cladophorales.

Thus, the chromosomes are organised at a metaphase plate which does not exceed the initial size of the resting nucleus (contrast *Cladophora*). Also the general appearance of the plate, the absence of centromeres and the size of the chromosomes are more similar to the Ulotrichales than the Cladophorales.

Sexual reproduction of Sphaeroplea

In most species the method is oogamous, but others show advanced anisogamy. At the onset of antherozoid formation the number of nuclei and chloroplasts in a cell increases in number. Progressive cleavage of the cytoplasm produces a number of uninucleate protoplasts, and each metamorphoses into a naked, spindle-shaped, biflagellate antherozoid. The cytoplasm of cells developing into oogonia divides into a number of egg cells without increase in the number of nuclei. The number of eggs is variable, even in the same species, and the antherozoids enter the oogonial cells through a number of pores. Thick-walled zygotes develop; they are liberated and normally undergo a resting period before germination.

Heterotrichous filamentous forms (Chaetophorales)

These forms are sometimes included as a separate family (Chaetophoraceae) in the Ulotrichales. Although this is supported by cytological evidence[1] (note, however, *Coleochaete*, p. 57), most authorities consider a heterotrichous type of construction warrants a distinct order.

Stigeoclonium (fig. 9 A) is a genus with many of the more simple features of the group and can be used as a point of reference of all others. Both the prostrate and erect systems are well developed; the filaments of the prostrate system are usually packed together to give a pseudoparenchymatous appearance, and branches of the erect system terminate in long hyaline hairs; the erect system shows diffuse growth whereas growth of the prostrate system is apical; there is no difference in size of cells between the main branches and the finer ones. Although the above features are general for the genus *Stigeoclonium*, the degree of branching, and the ratio of prostrate to erect system, are variable, and identification of species is difficult. Recent examinations of several species[1] suggest that chromosome numbers may provide an additional criterion.

Stigeoclonium reproduces asexually by the production of quadriflagellate zoospores, although in some instances smaller microzoospores are formed. Also, sexual reproduction has been

identified as isogamous fusion between biflagellate or quadriflagellate gametes. After a resting period the zygote usually germinates by the production of 4 zoospores, as in *Ulothrix*. However, on rare occasions a small unbranched filament grows out
from the zygote, and, since the nuclear divisions during development appear to be mitotic, this small filamentous body represents
a diploid stage. This stage later produces haploid zoospores.

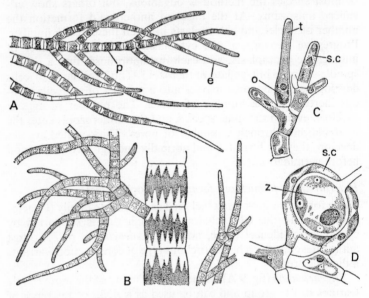

Fig. 9 Chaetophorales and Prasiolales

A, *Stigeoclonium lubricum*. B, *Draparnaldia glomerata*. C, *Coleochaete pulvinata*, young oogonium with trichogyne. D, *C. pulvinata*,
young sphermocarp. c, chloroplast; e, erect system with terminal
hair cell; o, oogonium with trichogyne (t); p, prostrate system;
s.c, cells forming (or which will form) ensheathing layer; z, zygote.
(B after Fott[84]; A, C, D after Smith[222])

Structural modifications of the heterotrichous plant body

Draparnaldia shows an important morphological elaboration;
the cells of the main branches are much larger than those of the
finer branches (fig. 9 B). This is also shown by *Draparnaldiopsis*,
which has the additional characteristic that cells of the main
branches consist of long and short cells and the laterals of limited

growth arise only from the short cells, thus showing a differentiation into nodes and internodes.

In many genera one or other of the systems is reduced so that it is sometimes difficult to identify the heterotrichous nature of the Algae. For example, in *Aphanochaete* the erect system is present only as a few hairs, whereas it is completely absent from *Endoderma*. Forms with only the prostrate system commonly occur as epiphytes and endophytes. Other genera (e.g. *Microthamnion*) consist only of the erect system. The reduction is carried even further in *Pleurococcus* which normally occurs either as single cells or small groups of 2–4 cells. No zoospores or gametes are formed and the cells merely reproduce by division of the vegetative cells. Occasionally the cells are arranged in the form of small branched filaments, and such forms, however rare, preclude it from inclusion in any other order.

Production of the reproductive bodies in specialised sporangia

Apart from *Endoderma*, where the zoospores are produced in enlarged cells clearly functioning as sporangia, most genera mentioned above form spores in normal vegetative cells. Species of Trentepohliaceae, however, have elaborate sporangia, and *Trentepohlia* has at least two kinds. The simplest of these are the *sessile* sporangia consisting of an enlarged vegetative cell occurring either in an intercalary position or terminally; the swarmers are liberated while the sporangia are attached to the plant. More elaborate *stalked sporangia* are formed by swelling of the apical part of a specialised outgrowth. Later the sporangium is released from the plant and quadriflagellate zoospores are produced.

Advanced oogamous sexual reproduction

Apart from some genera reproducing by an anisogamous process (e.g. *Aphanochaete*) most genera in the Chaetophorales show isogamy. *Coleochaete* is an exception; it has an advanced oogamous process (its chromosome number is also different from other members of the order[1]). Oogania are formed by an enlargement of the vegetative cell, and the enlarged cell develops a papilla which in *C. pulvinata* elongates into a prominent neck, a *trichogyne* (fig. 9 C). The antheridia are much smaller and are normally borne in clusters at the ends of branches and each antheridium liberates a small biflagellate colourless spermatozoid. These fertilise the ovum after entering through the papilla or trichogyne. The zygote remains in the oogonium, it increases in size and the wall thickens;

growth of branches from neighbouring cells forms an enclosing pseudoparenchymatous layer surrounding the oogonium. The combined oogonium and ensheathing material is termed a *spermocarp*. Germination of the zygote begins with meiotic nuclear division and the division continues until 8–32 daughter protoplasts are formed, each of which develops into a biflagellate zoospore.

Parenchymatous forms: 1. Ulvales

Apart from a few oblique or longitudinal divisions (e.g. *Cylindrocapsa*) all filamentous species divide only in the transverse plane so that the filaments are uniseriate. Members of the Ulvales, however, exhibit cell division in more than one plane so that a parenchymatous type of construction is formed (note some authorities do not separate such forms from the Ulotrichales since the young stages are uniseriate filaments closely resembling *Ulothrix*).

Schizomeris shows a simple type of parenchymatous thallus. In the early stages it is indistinguishable from *Ulothrix* but after some time longitudinal divisions occur and a narrow parenchymatous thallus is formed. More elaborate thalli are produced in *Ulva*, *Enteromorpha* and *Monostroma*. The young plant of *Ulva* appears as a uniseriate Ulotrichacean filament, but at an early stage cells divide longitudinally and one plane of longitudinal division is at right angles to all others, so that a flat 2-layered expanse of cells is produced. The early stages of *Enteromorpha* development are similar to those of *Ulva*, but later the 2 layers separate and each layer develops as a tubular structure. In *Monostroma*, longitudinal division begins at the earliest stage and no uniseriate filamentous stage can be identified; the flat expanse has one layer. Growth of *Ulva* is diffuse whereas that of *Enteromorpha* is partially apical. Vegetative propagation by fragmentation is common and all genera produce quadriflagellate zoospores and biflagellate gametes which fuse isogamously.

One interesting feature of *Ulva* and *Enteromorpha* is that isomorphic alternation of generations has been observed between a diploid sporophytic stage (meiosis occurs at zoospore formation) and a haploid gametophytic stage.

Parenchymatous forms: 2. Prasiolales

This group consists of algae that are either simply parenchymatous (small flattened blades) or little more than filamentous. They are distinguishable from the Ulvales by two main features: cell structure, and the form of life-history. When examined under the microscope the thallus appears as a flat sheet of cells, and the cells are small, polygonal and commonly arranged in groups of four. A distinctive feature is the large axile stellate chloroplast with a central pyrenoid.

Asexual reproduction occurs by the formation of akinetes formed from marginal cells. In 1960, a sexual process was identified in *Prasiola stipitata* and a unique kind of life-history has been elucidated.[88, 89] The young vegetative thallus develops either into a spore-forming plant or a gamete-forming plant. The former is diploid and reproduces asexually by the production of spores. In the gamete-forming plant the lower part is also diploid and is indistinguishable from the spore-forming plant. However, reduction division of some cells produces a mass of haploid tissue, some cells of which produce male gametes and others female gametes. A similar life-history has also been observed for *Prasiola meridionalis*.[31]

Oedogoniales

This order contains three genera: *Oedogonium*, *Oedocladium* and *Bulbochaete*. All are filamentous but there are two features which distinguish this order from all others; the method of vegetative cell division and the methods of asexual and sexual reproduction. The main features are common to all three genera, but the details below are taken from the most common genus, *Oedogonium*.

Vegetative cell division

The cells are uninucleate and normally have no vacuole. They have a well-defined wall, and a peripheral reticulate chloroplast with a pyrenoid. A series of fine parallel striations run across the distal end of the cell wall (fig. 10 F). Cells with these markings ('caps') are known as *cap cells* and they arise as a result of the method of cell division (figs. 10 A–D).

At the onset of cell division, the nucleus migrates from about the middle of the cell towards the distal end of the cell and divides mitotically. During the early stages of mitosis the inner layer of cell wall near the distal end of the cell thickens so that a complete ring of thickened wall material is formed, and this later becomes grooved. Thus at this stage the ring consists of a U-shaped thickened inner layer of the wall, with the open part of the U-shape covered by the outer layer (fig. 10 B). This outer layer finally ruptures so that the thickened portion becomes pulled out, and the cell therefore elongates to about double its normal length (fig. 10 C) with the distal half composed almost entirely of new wall material. Cytoplasmic cleavage follows nuclear division, and originally the septum occurs about mid-way along the original cell (fig. 10 B), but later coincides with the point where the new material begins (fig. 10 D). The remains of the original wall material near the distal end of the cell appears as a ring of thickened wall, and successive divisions produce a series of rings.

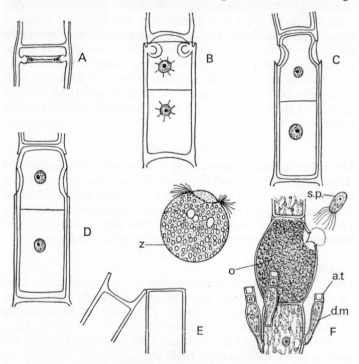

Fig. 10 Oedogoniales, *Oedogonium sp*

A-D, stages in cell division. E, zoospore being liberated. F, sexual reproduction. as, androsporangium (antheridium in macrandrous species); at, antheridium; d.m, dwarf male plant; o, oogonium; sp, spermatozoid; z, zoospore. (A–E after Fritsch[90]; F after Smith[222])

Asexual reproduction

Oedogonium reproduces asexually by the production of zoospores.

These are formed singly in vegetative cells and are released by rupture of the cell wall near the caps. Flagellation of the zoospore is characteristic; the anterior pole is surrounded by a ring of flagella (fig. 10 E), and fibrous strands connect the basal granules as a complete ring.[118, 119, 120] After swimming for about one hour they settle on a suitable substratum and germinate. One end of the zoospore normally functions as some kind of attaching organ, the other elongates slightly and division then begins.

Sexual reproduction

In the *Oedogoniales* this is a distinctive advanced type of oogamy; some species are homothallic and some heterothallic. The *oogonium* is formed by enlargement of a cap cell; it becomes spherical and approximately double the size of the vegetative cell, and becomes bright orange in colour (fig. 10 F). A small colourless area appears at one point on the side of the oogonium and fertilisation of the single ovum occurs through this. Male gametes (antherozoids), having the same structure as the zoospores, are formed singly, or in pairs, in antheridia. The formation of antheridia can occur in one of two ways. In some species small flattened antheridia are formed by successive rapid divisions of a cap cell. These species are termed *macrandrous* and contrast with the *nannandrous* species in which the swarmer released from an antheridial-like cell does not function as an antherozoid. It is termed an *androspore* and is liberated from an *androsporangium*. Although androspores swim towards the oogonium, they do not fuse with it. Instead, they become attached either to the oogonial wall or to a cell nearby (usually the cell immediately beneath it), and there germinate into small two- or three-celled filaments (fig. 10 F). The distal end of the filament functions as the antheridium (occasionally successive transverse divisions produce several small antheridia). The small filamentous structure is termed the *dwarf male plant*.

The zygote develops a two- or three-layered wall, and, after liberation by disintegration of the oogonial wall, it undergoes a resting period before germinating by the production of 4 zoospores.

Zygnematales

This order contains both unicellular and filamentous forms, and there are three important features which make this group different from other green algae. First, they show a marked elaboration of the chloroplast; secondly, flagellate stages are completely absent;

and thirdly, the algae reproduce sexually by fusion of amoeboid gametes during a specialised conjugation process. On the basis of these differences, Round[208] suggests that the members should be separated from other green algae in a distinct class, Conjugatophyceae.

Filamentous forms are included in the family, *Zygnemataceae*, and the unicellular species are divided between two families, *Mesotaeniaceae* and *Desmidaceae*. It is convenient to present the filamentous forms first since they illustrate most of the essential features of the group, and the unicellular species can be related to these.

Zygnemataceae

The filaments are usually unbranched and free-floating. The cylindrical cells show great variety of chloroplast form and the character is the basis of classification of the family.

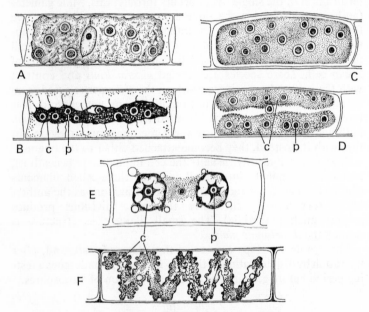

Fig. 11 Zygnematales, chloroplast form

A, B, *Mougeotia sp.* A, face-view; B, side-view. C, D, *Sirocladium kumaolense*: C, face-view; D, side-view. E, *Zygnema terreste*. F, *Spirogyra sp.* c, chloroplast; p, pyrenoid. (After Randhawa[203])

(a) *Single axile plate-like chloroplasts* occur towards the centre of the cells of *Mougeotia, Mougeotiopsis, Temuogametum,* etc. (fig. 11 A, B). *Mougeotia* shows an interesting orientation of the chloroplast with changes in the direction of illumination and rotates so that its narrow face is to the direction of bright light. Phytochrome is the pigment responsible for light perception and the phytochrome molecules are arranged in a helical pattern in the outermost layer of the protoplasm.[111]

(b) *Two parietal plate-like chloroplasts* are present in the genus *Sirocladium.* The plates are located near the surface of the cell and, when viewed with the full face of the chloroplast in view, only one can be seen but in side-view two can be seen (fig. 11 C, D).

(c) *Stellate axile chloroplasts* occur in *Zygnema* (two in each cell) (fig. 11 E). In other genera (e.g. *Zygogonium*) the chloroplasts tend to be more rounded and are not so regularly stellate.

(d) *Helical chloroplasts* are found in two genera, *Sirogonium* and *Spirogyra* (fig. 11 F). The chloroplast of *Spirogyra* is either narrow with an entire margin, or is broad with a serrated edge and runs in an anticlockwise direction around the cell. In *Sirogonium* the bands run in a longitudinal direction.

Although chloroplast form has been studied most extensively other features of cell structure are of interest. The cell wall is two-layered with an inner cellulose layer and an outer one of pectic material. At the time of septum formation, a septum of pectic material is formed first and a layer of cellulose is later formed each side of it; thus a 'middle lamella', superficially like that in higher plants, is formed.

The central region of the cell is usually occupied by a large vacuole (this is modified in species with axile chloroplasts) and the vacuolar sap is frequently rich in tannin material (in *Zygogonium ericetorum* and *Zygnema terrestre* the sap is coloured purple by the pigment *phycoporphyrin*).

The nucleus is usually in the centre of the cell in a band of cytoplasm which traverses the central vacuole. This is not always so, since in *Mougeotia* it is closely adpressed to the chloroplast. Large complex nucleoli are present and considerable lengths of the nucleolar organising chromosomes are involved in the organisation of the nucleolus. Before nuclear division the nucleoli disperse. This process is variable in both time and degree, and this might account for the controversy in the earlier literature over the role of the nucleolus.[100]

Reproduction

Vegetative propagation by fragmentation is common, since the 'middle lamella' is sometimes sensitive to breakages. Akinetes and aplanospores are sometimes formed.

Sexual reproduction is achieved by fusion of amoeboid gametes, a process known as *conjugation*. There are two kinds of conjugation, *lateral* and *scalariform*. The former involves fusion of gametes derived from cells of the same filament and the latter between cells of different filaments. When the gametes fuse in the conjugation tube the process is isogamous, but when only one moves to fuse with the other in the parent cell, it is anisogamous. Species showing lateral conjugation include *Zygogonium indicum* and several species of *Zygnema* and *Spirogyra*. Scalariform is much more common; it is isogamous in species of *Debarya* and *Mougeotia*, and anisogamous in species of *Zygnema* and *Spirogyra*. In the anisogamous processes the female gametangia, and hence the zygospores, sometimes all occur in the same filament, and in other species 'male' and 'female' gametangia are mixed on the same filament. The details of the conjugation process have been established for several species of *Spirogyra* and many other genera appear to be similar. At the onset of conjugation the filaments become attached to each other along their lengths, with the production of much mucilage. Papillae are then formed, first from one filament and later from the other, so that the two filaments are pushed apart. The wall between the papillae breaks down and a conjugation tube is formed; the male protoplast contracts and is forced (probably the result of surface tension effects) through the conjugation tube to fuse with the female gamete. Nuclear fusion normally occurs immediately after cytoplasmic fusion. The zygospore remains dormant for some time, and during ripening, meiotic nuclear division occurs and three of the four daughter nuclei abort. At germination, the thickened wall ruptures and the protoplast (or part of it) is extruded; transverse divisions lead to the formation of the filamentous plant.

Unicellular forms (*Mesotaeniaceae and Desmidiaceae*)

Organisms of these two groups are predominantly unicellular, although in both families a few short filaments are known. The two groups are distinguished from each other by two main features: 1. The cells of the *Desmidiaceae* ('placoderm' or 'true' desmids) are constricted to show two semicells and the wall is in two pieces, whereas in the *Mesotaeniaceae* (saccoderm desmids) the wall is composed of one piece with no constriction into 2 semicells. 2. The cell wall of the true desmids has 2–3 layers and has a number of pores, whereas those of the Mesotaeniaceae have a single-layered wall with no pores.

The Mesotaeniaceae show obvious affinities with the Zygnemataceae, since they show a similar range of chloroplast form. For example, the chloroplast of *Mesotaenium* is a flat axile plate with one or several pyrenoids; *Cylindrocystis* has a pair of stellate axile chloroplasts with a single large pyrenoid at the centre of each chloroplast; and the chloroplast of *Spirotaenia* is a helical band with a number of scattered pyrenoids. Also the mode of cell division in Mesotaeniaceae resembles that of the Zygnemataceae with the formation of a 'middle lamella' and deposition of the cell wall on each side, but unlike the filamentous forms the cells separate immediately by disintegration of the 'middle lamella'.

Asexual reproduction is uncommon, and sexual fusion occurs by the fusion of amoeboid gametes. At the onset of conjugation, two cells embedded in mucilage become joined by a conjugation tube formed by the meeting and fusing of papillae as described for filamentous forms. The gametes normally fuse in the middle of the conjugation tube but the nuclei do not usually fuse until after the zygospore ripens. Germination of the zygote begins with reduction division of the nucleus and 4 new cells are formed around the 4 nuclei (contrast the 'true' desmids below).

The Desmidiaceae is a much larger group of organisms and the diversity of cell shape is great; some common forms are shown in fig. 12.

Although species show a high degree of specialisation of chloroplast form it is not possible to identify the three types found in the two families above, and an axile chloroplast in each semicell is most common. The cell wall is composed of two halves and possesses many pores, through which mucilage appears to be extruded. It has been suggested that localised exudation of mucilage might be related to the gliding of desmids.[127]

In unconstricted species such as *Hyalotheca* the mode of cell division resembles that of the Mesotaeniaceae, but the process is complicated in those species constricted into two semicells. Most commonly, the *isthmus* (that is, the part of the cell connecting the two semicells) elongates so that the 2 semicells are forced apart. A membrane, and later a wall, develops across the isthmus, the two cells separate and the small part of each (that is, the remains

C

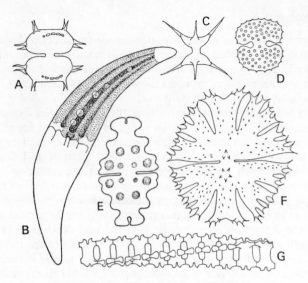

Fig. 12 Vegetative cells of some desmids

A, *Xanthidium antilopaeum* var *polymazum*. B, *Closterium monili-ferum*. C, *Staurastrum curvatum*. D, *Cosmarium reniforme*. E, *Euastrum affine*. F, *Micrasterias apiculata*. G, *Desmidium aptogonium*. (After Smith[222])

of the isthmus) enlarges to form a new semicell. Thus one semicell consists of old cell wall material and the other of new.

Asexual reproduction is rare, but sexual reproduction occurs between conjugating cells embedded in mucilage. In many species, the semicells separate, protoplasts are liberated and fuse (that is, no conjugation tube is formed). In species of *Closterium* and *Desmidium* protuberances develop from each isthmus, a conjugation tube is formed and the gametes normally fuse in this tube (the anisogamous process in *Desmidium cylindricum* is an exception). Nuclear fusion does not occur until immediately before germination. Germination begins with reduction division of the nucleus; two of the four nuclei degenerate and the remaining two develop into new individuals.

MULTINUCLEATE FORMS (CLADOPHORALES, ACROSIPHONALES, SIPHONOCLADALES, DASYCLADALES AND CAULERPALES)

The classification of multinucleate forms of green algae is variable and depends on the opinion of the particular authority. For the present discussion such forms are to be classified into 5 orders, and any modifications will be discussed at the appropriate time.

Cladophorales

Organisms in this order are multicellular and the filaments may be branched (*Cladophora*) or unbranched (*Chaetomorpha*). The multinucleate cells are elongated; the walls are characteristically thickened and stratified with an inner cellulose layer, a middle layer of pectic material and an outer insoluble layer possibly composed of chitinous material. The cytoplasm occupies a peripheral lining layer containing a reticulate chloroplast, and the bulk of the cell is occupied by a large central vacuole.

Each cell is generally capable of division, although in genera such as *Cladophora* and *Pithophora* division is partly apical. Cell division begins with a bar of thickening running round the cell, and the thickened area later grows centripetally to cut across the cell lumen. The process of cell division is not related cytologically to nuclear division.

Asexual reproduction is achieved by the formation of quadriflagellate zoospores in actively growing parts of the filament. Nuclear division preceding zoospore formation is meiotic. Sexual reproduction occurs by fusion of identical biflagellate gametes formed in the same way as zoospores.

An isomorphic alternation of generations has been established in species of *Cladophora*: the haploid stage produces biflagellate gametes, germination of the zygote produces an identical diploid stage which reproduces asexually by the production of haploid zoospores.

Acrosiphonales

This order contains the genera *Acrosiphonia*, *Spongomorpha* and *Urospora*.[191] All of these were, at one time, included in the Cladophorales and they have many features in common with this latter group. However, two characteristics argue for their separation as a distinct order:

1. *Cell wall composition*. Whereas Cellulose I is found in the walls of Cladophorales, it is replaced by Cellulose II (mercerised cellulose) in those genera now included in the Acrosiphonales.[180, 181]

2. *Life-history*. The haploid and diploid stages of *Cladophora* are morphologically similar; this is not so for members of the

Acrosiphonales.[76, 128, 129, 130] *Spongomorpha coalita, Urospora mirabilis* and *Acrosiphonia spinescens* reproduce solely by isogamous fusion of the gametes and the zygote germinales into a form previously described as *Codiolum* (an ovoid multinucleate vesicle with a stalk). *Spongomorpha lanosa* also shows an heteromorphic life-history in which the zygote germinates into a form resembling *Chlorochytrium inclusum*.

The use of a heteromorphic life-history to create a separate order, the *Acrosiphonales*, is cast in doubt by the observation[6] that under certain conditions *Codiolum*-like plants are formed in cultures of *Cladophora rupestris*. This is not, however, the usual form of life-history, and the observation has not been repeated.

Siphonocladales

This order also includes multicellular species in which the cells are multinucleate and possess a parietal reticulate chloroplast. They have a characteristic process of *segregative cell division* in which cytoplasmic cleavage separates a number of multinucleate fragments of the protoplast from the remainder of the cell; a wall is formed later and new coenocytic vegetative cells are formed.

Valonia begins as a spherical cell which later acquires a large central vacuole and the mature plant therefore appears as a balloon-shaped vesicle anchored to the substratum by several unicellular rhizoids. Formation of biflagellate swarmers has been observed to be preceded by meiotic nuclear division, fusion is isogamous and the zygote germinates into a new vegetative plant. Other genera are more elaborate. For example, *Siphonocladus* begins as a small, erect, balloon-shaped cell, the contents of which later segregate into a large number of protoplasmic units; each of these grows through the original cell wall to form a branch. Thus the mature thallus consists of a central globular area with many projecting branches.

A point of taxonomic interest is whether Siphonocladales should be isolated from the Cladophorales. The cell wall structure of *Valonia* and *Cladophora* are similar, and because of this, Chapman[22, 23] puts them in the same order. Most people's (e.g. Egerod's)[72] objections to their being in the same order are based on the absence of an initial vesicle stage from the Cladophorales, (although Chihara[25] has observed such a stage in *Cladophora wrightiana*) their cytology is also different.

Dasycladales

The organisms of this order exist in two forms. In all genera the juvenile stage consists of an undifferentiated rhizoidal system with a small erect axis showing slight development of whorled structures at the apex. This vegetative stage is usually uninucleate. The reproductive stage, on the other hand, consists of an erect axis with whorls of branches near the apex.

A widely known genus is *Acetabularia*: it has been used for intensive investigation into many biochemical aspects of morphogenesis.[108] In this genus most of the whorled structures are gametangia, and the vegetative part is reduced to a minimum. This represents a distinct increase in specialisation, since *Dasycladus*, for example, has an extensive lateral system with gametangia produced only at the tips of certain branches.

When gametangia are formed, the nucleus divides mitotically and the large number of nuclei so produced migrate to the gametangia and cytoplasmic cleavage produces a number of uninucleate cysts. The nucleus divides (meiosis has been observed at this stage) and a number of biflagellate gametes are finally liberated. Similar gametes fuse, and the zygote germinates into the vegetative rhizoidal thallus.

Members of the Dasycladales frequently become encrusted with lime. Because of this, there is an interesting fossil record of the development within the group. In the Carboniferous the erect axis has lateral appendages irregularly arranged with no appearance of a whorled structure until the Triassic.

Caulerpales

Plants of this group are coenocytic and are frequently extremely large and complex. The order has four main features different from all other Chlorophyceae: 1. α-carotene replaces β-carotene, 2. the vegetative plant is diploid, 3. there are many discoid chloroplasts, 4. all are marine.

Halicystis is a convenient starting point for consideration of the group since it shows many of the simplest features. It is a small green vesicle containing numerous nuclei and chloroplasts, and is attached to the substratum by means of a rhizoid. Thus it bears a superficial resemblance to *Protosiphon* (p. 48) and it might be suggested that *Protosiphon* should be included in the Caulerpales. There are three main reasons for rejecting this suggestion: 1. *Protosiphon* has β-carotene; 2. it is haploid and reduction division occurs during germination of the zygote and not during gamete formation; 3. it has a single reticulate chloroplast instead of many discoid ones.

Other genera are more elaborate than *Halicystis*, and the following genera illustrate the more important developments:

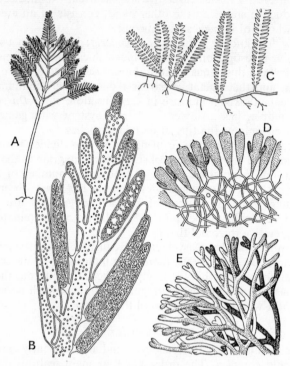

Fig. 13 Caulerpales

A, B, *Bryopsis corticulans*: A, habit; B, apex with gametangia. C, *Caulerpa crassifolia*. D, E, *Codium fragile*: D, transverse section of a thallus branch; E, habit. (After Smith[222])

(a) *Bryopsis* has a rhizoidal portion with erect system showing extremely regular branching in a precise pinnater or bipinnate manner (fig. 13 A, B).

(b) *Caulerpa* has a creeping rhizome-like portion from which small rhizoidal-like structures arise on the lower side and erect branches of complex and variable form arise from the upper side (fig. 13 C). There are two main methods of strengthening the coenocytic thallus, encrusting with lime or interwining of the filaments. Neither is used by *Caulerpa*; instead the central vacuole is traversed by a number of projections from the cell wall.

(c) *Codium* is a genus with a large elaborate pseudoparenchymatous thallus strengthened by interwining of the filaments and tight packing of *utricles* (fig. 13 D, E). There is usually differentiation between the central *medulla* (consisting of axial threads frequently having no chloroplasts) and the cortex. The axial threads turn at right angles, and the tips swell to form the utricles of the cortex.

Asexual reproduction is rare or absent from the Caulerpales, but a sexual process is common. In *Halicystis* the process is anisogamous between two biflagellate uninucleate gametes; the female gamete is larger with more chloroplasts than the male. The gametes are produced by different plants and are formed in special gametangial areas separated from the rest of the protoplast by a membrane. Fusion occurs in the open sea (the release of gametes shows marked periodicity) and the zygote germinates into a tubular outgrowth described under the name of *Derbesia marina*. The zoospores formed by this plant have a ring of flagella (cf. *Oedogonium*) and they germinate into the *Halicystis* plant. Although an alternation has been established between the *Halicystis* and *Derbesia* forms, the location of meiosis is unknown.

Anisogamy also occurs in *Bryopsis* and *Codium*. In the former gametangia are formed at the tips of ultimate branches, and the male and female gametes are produced by separate plants. In *Codium*, the gametangia are formed as small branches growing from the sides of the utricles.

Dichotomosiphon is an exception since it has an oogamous method of sexual reproduction. The sex organs are borne at the tips of short curved branches of the coenocytic filament. A septum cuts off the terminal part of the branch as the antheridium and cleavage of the cytoplasm produces many small biflagellate antherozoids which are liberated through a pore. The oogonium is a large spherical structure formed by swelling of a tip of a branch; it is cut off by a septum and is fertilised through a papilla.

5

CHLOROPHYTA: CHAROPHYCEAE

Division CHLOROPHYTA
Class CHAROPHYCEAE

The Charophyceae includes 8 genera, all of which show a complex morphology and a degree of specialisation of the sexual reproductive process greater than is found in any of the Chlorophyceae. The class differs from the Chlorophyceae in four important ways. 1. The plant body consists of an erect axis differentiated into nodes and internodes, with laterals of limited growth at the nodes. 2. The reproductive organs are complex and are surrounded by a zone of sterile tissue. 3. The motile male gamete, antherozoid, is unlike any found in the Chlorophyceae, being an elongated biflagellate spiral structure. 4. The zygote germinates into a protonemal stage, from which the adult plant develops later.

The two most common genera in the Charophyceae are *Chara* and *Nitella* and the discussion below is concerned almost entirely with these two genera.

VEGETATIVE STRUCTURE

The thallus consists of an erect branched axis attached to the substratum by an elaborate system of multicellular rhizoids. The erect system has well-defined nodes and internodes. In both *Chara* and *Nitella* the nodes consist of several small isodiametric cells in a cluster of two central cells surrounded by 6–20 peripheral cells. In *Nitella* the internode consists of a single elongated cell, whereas in *Chara* the single central internodal cell is surrounded by a cortex of smaller elongated cells, half of which are derived from the node above and half from the node below.

A whorl of short laterals arises from each node and each lateral develops from one of the peripheral cells of the node. These laterals grow to a length characteristic of the particular genus or species and growth then ceases. The number, degree of branching and the detailed structure of the laterals are important diagnostic features in establishing genus and species. For example, the laterals of *Nitella* are simple or branched uniseriate filaments, whereas in *Chara* the structure of the laterals resembles that of the main axis.

Growth of the Charophyceae is by a single, large apical cell located at the tip of the axis. When it divides a segment is cut off the posterior, or bottom face of the cell, and the segment divides again by a transverse wall. As a result of this a lower internodal cell is formed. This internodal cell does not divide again but elongates to the length of the mature internode. The upper cell divides a number of times to form the nodal elements.

DETAILS OF CELL STRUCTURE

In the smaller cells of the node the cytoplasm is dense, a single nucleus is located towards the centre of the cell and a number of small discoid chloroplasts devoid of pyrenoids are distributed throughout the cytoplasm. In the large internodal cell the central region is occupied chiefly by a large central vacuole, with the small chloroplasts arranged in well-defined longitudinal series in the peripheral cytoplasm. The nucleus (sometimes lobed) is also located in the lining cytoplasm, and later it divides amitotically to form a large number of nuclei in the lining layer.

The lining layer of cytoplasm is made up of two distinct areas, a stationary *exoplasm*, and an *endoplasm* which undergoes constant rotation. Since the cytoplasm flows down one side of the cell and up the other, there must be two stationary strips where the pathways adjoin, and these strips are characterised by a special structure of the wall and probably of the cytoplasm also.[131] Unlike the chloroplasts of higher plants the chloroplasts do not circulate in the endoplasm but are maintained in the stationary exoplasm. Jarosch[125, 126] has suggested a possible mechanism for this cytoplasmic streaming in the cells of the Charophyceae. He suggests that there are protein fibrils of the cytoplasm fixed to the cell wall, and that these fibrils undergo regular contraction and expansion so that a transverse wave passes along the fibril.

Because the fibril is fixed to the wall the passage of such a wave causes the cytoplasm adjacent to the fibril (that is, the endoplasm) to stream. Although there is no definitive evidence for this hypothesis, fibrils of a wavy nature have been observed in the cytoplasm of the cells of the Charophyceae.[131] It is this scheme of cytoplasmic streaming which formed the basis for Jarosch's hypothesis of the mechanism of gliding in blue-green algae (see p. 31).

REPRODUCTION

Apart from vegetative propagation by the formation of a number of outgrowths from the plant, the only method of reproduction is an elaborate oogamous sexual process. The reproductive structures are borne at the nodes and their detailed arrangement (that is, whether on separate plants, or, if on the same plant, the spatial arrangement of the male in relation to the female) is dependent on species.

The mature male structure is termed a *globule*, and the antherozoids are borne singly in antheridia which are contained within the globule. The antheridia occur as delicate antheridial filaments radiating from a mass of cells and the entire globular structure is surrounded by a single layer of cells (fig. 14 J). Intermingled with the antheridial filaments are filaments of vegetative cells radiating from the central cells and extending to the outer layer of cells.

The mature female structure is the *nucule* and consists of a central oogonium surrounded by spirally arranged cortical cells (fig. 14 E).

Thus each reproductive structure is a complex structure consisting of the actual gamete-bearing cells surrounded by sterile tissue. They therefore bear a strong superficial resemblance to the multicellular structures of archegoniate plants and at first sight appear to invalidate the main criterion by which an alga is distinguished from an archegoniate plant (see p. 9). It is therefore important to establish whether the sterile tissue surrounding the reproductive structures is to be interpreted as part of the reproductive body or as modified vegetative tissue. The correct interpretation depends on an examination of the ontogeny of each structure.

The formation of both structures begins with divisions of a superficial cell of the node, and one of the daughter cells produced by this division functions as the initial. In the development of the female structure, the initial divides to form a row of three cells, the lowest one of which functions as the pedical of the structure and does not divide further. The terminal cell is the *oogonial mother cell* and divides to give a lower *stalk cell* and a terminal *oogonium* (fig. 14 D). The middle cell of the original row of three

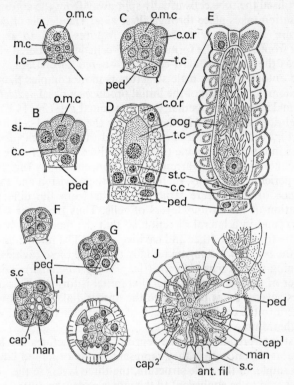

Fig. 14 Charophyceae, *Chara sp.*

Successive stages in development of a nucule (A-E) and a globule (F-J). ant.fil, antheridial filament; c.c, central cell; cap [1], primary capitulum; cap [2], secondary capitulum; cor, corona; l.c, lower cell; m.c, medium cell; man, manubrium; o.m.c, oogonial mother cell; oog, oogonium; ped, pedicel; s.c, sheath cell; s.i, sheath initial; st.c, stalk cell; t.c, tube cell. (After Smith[222])

divides vertically a number of times to give a central cell surrounded by 5 *sheath cells*. These latter cells grow up to surround the oogonium. When the oogonium elongates, the cells also enlarge and become spirally arranged around the mature oogonium (fig. 14 E). Before elongation the sheath cells divide transversely and only the lower cells (the *tube cells*) elongate; the upper cells do not enlarge significantly and they constitute the *corona* of the mature structure (fig. 14 D, E). When the egg cell is ready to be fertilised the tube cells usually split away from the corona and fertilisation takes place through these slits. The details of corona structure are important taxonomic features used to separate *Chara* from *Nitella*; the former has a 5-celled single-tiered corona whereas the latter has a 10–celled two-tiered corona.

The ontogeny of the male structure is more complex than that of the female. Division of the initial produces a basal *pedicel* as in the female structure together with a terminal cell (fig. 14 F). The pedicel does not divide further whereas the terminal cell divides three times to produce a compact octad of cells, each of which divides in a periclinal plane to produce three layers of cells (each layer containing 8 cells) arranged in a concentric way (fig. 14 G). After periclinal divisions, there is no further division and the appearance of the mature structure results from the differential elongation of the various layers of cells. Thus the outer (*shield*) cells expand in a lateral direction so that the inner cells become separate from each other and cavities result. The middle layer of cells (the *manubria*) elongate radially to bridge these cavities. The remaining central mass of cells (the *primary capitulum*) divides a number of times until an antheridial initial is formed. This initial divides several times to produce the antheridial filaments which radiate into the cavities. Each cell of these filaments functions as an antheridium.

Most interpretations of the reproductive structures are based on the assumption that they represent metamorphosed laterals. For example, in the male structure, the three layers in the octant are assumed to be equivalent to the upper node (the central cells), the internode (the middle layer) and the lower node (the outer later). On the basis of this interpretation, the antheridial filaments represent undifferentiated laterals arising from the upper node. Similarly the row of three cells formed during the early development of the female structure are assumed to represent modified internode (basal cell), node (middle cell) and internode (upper cell). The interpretation of the middle cell as a nodal one is based on the fact that it divides to form a central cell surrounded by a number of sheath cells. On the basis of this interpretation the reproductive organs (the antheridia and archegonia) themselves are actually unicellular and the surrounding sterile tissue is modified vegetative tissue. However, it is difficult to decide whether the above interpretation is truly objective, or whether it is proposed merely because it allows a sharp demarcation to be drawn between archegoniate plants and algae. Even if the interpretation is correct, the protection of the reproductive organs by sterile tissue (whether vegetative or not) is still, evolutionary speaking, an advance comparable to that found in the development of such a protection in archegoniate plants, and the differences between the reproductive structures of the Charophyceae and those of all other algae should be emphasised.

GERMINATION OF THE ZYGOTE

Before germination of the zygote the wall thickens, the nucleus migrates to the apical pole and there divides into four daughter nuclei. Although this formation of four nuclei suggests that the division is meiotic, there is no direct evidence, and Ross[206] has failed to confirm earlier suggestions of meiosis during the germination of the zygote of *Chara gymnopitys*. However, it appears most likely that the plant is haploid since no reduction division has been seen to accompany the formation of gametes. After formation of four nuclei, a small distal uninucleate cell is cut off from the major part of the zygote, and this larger part later degenerates. The small cell then divides vertically to form a rhizoidal initial and a protonemal initial. The former develops into a colourless rhizoid and the latter divides to form a small filamentous primary protonema. Both the rhizoids and the protonema are differentiated into nodes and internodes and one of the appendages borne on the second node of the protonema develops into the mature axis.

PHYLOGENY OF THE CHAROPHYCEAE

The Charophyceae can be divided into two sub-groups, the *Charoideae* and the *Nitelloideae*, the former having 5 coronal cells and the latter 10. These appear to have diverged early in the evolution of the group and the corticate forms are generally thought to have evolved from the ecorticate. Thus the usual outline of evolution of the Charophyceae presents the more primitive species as those of the Nitelloideae (that is, the genera *Tolypella* and *Nitella*), and those of the Charoideae are generally thought to be derived from these. Within the Charoideae the direction of evolution is usually assumed to be from the ecorticate to the corticate forms, that is the evolutionary sequence begins with the ecorticate *Protochara* and passes through *Nitellopsis*, *Lampro-*

thamnion, Lychnothamnus to the extensively corticated *Chara*.
Desikachary and Sundaralingam[43] have suggested that evolution
of the Charophyceae has occurred in precisely the opposite direction from the above scheme; they suggest that *Nitella* is more
advanced than *Chara* and that within the Charoideae the ecorticate forms have arisen from the corticate. The main reasons for
their suggestions are that *Nitella* has sympodial branching of its
laterals; also, there is a greater amount of sterile tissue in the
female reproductive organ. Moreover these authors suggest that
on the basis of this approach to the evolution of the Charophyceae it is possible to devise a scheme for the origin of the Charophyceae from a corticated form related to *Draparnaldiopsis* (a
member of the Chaetophorales in the Chlorophyceae). *Draparnaldiopsis* has four main features which suggest affinity with the
Charophyceae: 1. it is differentiated into nodes and internodes,
2. it has a cortex-like covering derived from the basal cells of the
laterals, 3. the laterals have limited growth and 4. the reproductive organs are restricted to the laterals. However, there is no
fossil evidence to help decide between the two alternative
phylogenetic schemes for the Charophyceae.

6

XANTHOPHYTA: XANTHOPHYCEAE

Division XANTHOPHYTA
Class XANTHOPHYCEAE

The species of this class were at one time included in the Chlorophyceae. Because of their unequal flagella they were classified in the sub-group 'Heterokontae' and the others with equal flagella were classified as 'Isokontae'. As an increasing number of species was examined, the difference in flagellation was found to be paralleled by other important differences, and the separate class was created.

The Xanthophyceae are characterised by four main features. 1. There is frequently an excess of carotenoids over chlorophylls; moreover, chlorophyll *e* appears to be restricted to this division, although up to now it has been found in only two genera (see p. 11). 2. The two flagella are normally of unequal length, the longer being of the pantonematic type and the shorter acronematic. 3. The food storage products are normally oil and fat, but starch never appears to accumulate. 4. A cell wall is frequently absent and when present it generally has a higher content of pectic material than the cell walls of the Chlorophyta. Also, silicification of the cell wall is common and it frequently appears to be composed of two halves.

The class is much smaller than the Chlorophyceae and with a few exceptions all members are freshwater. Despite a reduction in the extent of morphological diversity, the class shows many developments which parallel those found in the Chlorophyceae.

Thus the primary division of the Xanthophyceae into its 6 orders is made using the vegetative structure as the major criterion, and on this basis five of the orders parallel those of the Chlorophyceae; the Heterochloridales (motile), Heterogloeales (palmelloid), Mischococcales (coccoid), Tribonematales (filamentous) and the Vaucheriales (siphoneous) being the counterparts of the Volvocales, Tetrasporales, Chlorococcales, Ulotrichales, and the Caulerpales, respectively, of the Chlorophyceae. The sixth order is the Rhizochloridales exhibiting a vegetative type not found in the Chlorophyceae, namely the amoeboid form.

Species of Xanthophyceae have been largely neglected in recent studies on the fine structure and life-histories of algae, so that many of the comments below may require modification when species are investigated with modern techniques.

The Heterochloridales includes those species in which the vegetative cell is motile, but unlike the Volvocales of the Chlorophyceae the motile colony is unknown, and all species are unicellular. Most of the species are naked, and in many the cytoplasmic membrane is not rigid, so that there is a tendency for cells to become amoeboid (e.g. *Heterochloris*).

No species is common but the main features of the order can best be illustrated by reference to the best-known genera, *Chloramoeba* and *Heterochloris*. For example, *Chloramoeba* shows the form of chloroplast most common in the Xanthophyceae, namely, 2–6 discoid chloroplasts located towards the periphery of the cell. In *Heterochloris*, the chloroplasts are also parietal but are only two in number and are located at the sides of the cell. The cells of *Chloromeson* have a single central chloroplast (fig. 15 A, B).

The normal reserve products are oil and fat, and because no starch is deposited one cannot resolve the question of whether the pyrenoid-like bodies are functionally equivalent in the pyrenoids of the Chlorophyceae.

Some species show a marked tendency to lose their photosynthetic pigments. For example, cells of *Chloramoeba* become colourless when grown under heterotrophic conditions, and although it has been stated (ref. 90, p. 471) that colourless cells have no chloroplasts, observations with the electron microscope have not been made to investigate whether leucoplasts are present as in *Polytoma* (p. 42).

A single contractile vacuole is normally present, but there appears to be no stigma. The cells resemble those of all Xanthophyceae (except the siphoneous forms) in possessing a single nucleus. One interesting feature is the presence of a single fibrous strand (rhizoplast) linking the base of the flagella to the nucleus. Such rhizoplasts have also been observed in the cells of the Chrysophyta, Pyrrophyta, Phaeophyta and Euglenophyta. Although the flagellar bases have sometimes been implicated in the nuclear division, extensive observations with the euglenoids

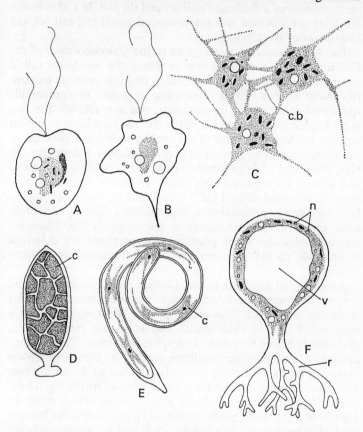

Fig. 15 Xanthophyceae

A, *Chloromeson agile*. B, *C. agile* with rhizopodia. C, *Chlorarachnion reptans*. D, *Characiopsis turgida*. E, *Ophiocytium variabile*. F, *Botrydium granulatum*. c, chloroplasts; c.b, cytoplasmic bridge; n, nuclei; r, rhizoid; v, vacuole. (A, B, F after Fott[84]; C after Smith[222]; D, E after Fritsch[90])

have not confirmed this (see p. 125); the same conclusion might also apply to the Xanthophyceae.

The normal method of reproduction of the motile unicells is by longitudinal division when the cell is still motile. In *Heterochloris* the products of division sometimes fail to separate and extensive palmelloid masses can result. Many genera also produce cysts by the formation of a thickened wall around the cell. In *Chloromeson* the cysts are silicified and are produced inside the cell (cf. the *statospores* of the Chrysophyta).

The *Mischococcales* is analogous to the Chlorococcales of the Chlorophyta. It contains forms in which the vegetative cell is non-motile. All species are unicellular, and it is possible to identify many species which show close analogies to comparable species of green algae. In particular, one can identify the two main tendencies, the trend towards the multinucleate condition and the tendency for some of the species to become attached to the substratum. The group includes *Botrydiopsis*, a common soil alga, consisting of large spherical cells with a pectic membrane of two halves, impregnated with silica. There are numerous peripherally arranged discoid chloroplasts and asexual reproduction is effected by the production of large numbers of zoospores, or, in some genera, these are replaced by aplanospores (e.g. *Chlorobotrys* and *Monodus* never produce zoospores and are therefore comparable to the azoosporic series of forms in the Chlorococcales).

Amongst the attached forms, *Characiopsis* is widespread and consists of elongated cells attached to the substratum by a short stalk (fig. 15 D) (cf. *Characium*, p. 47). Most species have a number of discoid chloroplasts and some are multinucleate, particularly before the onset of zoospore formation. The tendency towards a multinucleate condition reaches its climax in *Ophiocytium*, a genus with elongated cells attached to the substratum and which is multinucleate throughout most of its life (fig. 15 E).

The *Rhizochloridales* includes those species which have an amoeboid form. Some of the species of Heterochloridales become amoeboid temporarily (e.g. see fig. 15 B), but in the present order the species are permanently amoeboid, and apart from the formation of zoospores by the single multinucleate species, *Myxochloris*, the sole method of reproduction is by vegetative division of the cells. Many of the amoeboid forms have a tendency to become connected together by fine cytoplasmic strands, so that *Chlorarachnion*, for example, consists of a large mass of cells (about 150) joined in such a way (fig. 15 C). As in other amoeboid cells (e.g. those of the Chrysophyta) species of the Rhizochloridales are frequently holozoic.

In the *Heterogloeales* the vegetative state consists of a mass of non-motile cells embedded in mucilage. *Gloechloris* consists of spherical colonies with ellipsoideal cells irregularly arranged in a uniform gelatinous mass, and reproduction is normally achieved by the production of 2 naked biflagellate zoospores from each cell. The common freshwater genus, *Botryococcus*, is sometimes included in this group, but the biochemical evidence shows it to belong to the Chlorophyceae. For example, Belcher and Fogg[8] have confirmed the presence of starch and have also found chlorophylls *a* and *b*.

The *Tribonematales* include all multicellular filamentous genera, and most are unbranched (*Monocilia* sometimes forms branched filaments, but more commonly it appears as small packets of cells comparable to *Pleurococcus*). *Tribonema* is the only common genus and bears a strong superficial resemblance to the green alga, *Microspora* (p. 54). The cells of the unbranched filaments have walls composed of two halves, and when the filament breaks characteristic H-pieces are formed. Careful observation confirms that it is a member of the Xanthophyceae and not the Chlorophyceae since the storage products are usually deposited as oil and possibly chrysolaminarin. Moreover it reproduces by the production of 'typical' Xanthophycean biflagellate zoospores. Sexual reproduction has been observed and is apparently isogamous, although one gamete stops swimming and loses its flagella immediately before fusion.

The *Vaucheriales* is analogous to the Caulerpales of the Chlorophyceae, but is much more restricted. Only two genera, *Botrydium* and *Vaucheria*, are common and a discussion of these can illustrate the main features of the group.

Botrydium, a common alga growing on the surface of mud, bears a strong superficial resemblance to *Protosiphon* (p. 47) since it consists of an aerial vesicular part rooted in the mud by a branched system of colourless rhizoids (fig. 15 F). The vesicle part has a large central vacuole and a peripheral layer of cytoplasm containing numerous nuclei and discoid chloroplasts. Reproduction occurs by the formation and liberation of aplanospores or biflagellate swarmers. The latter appear generally to function as zoospores although there is one report that they behave as gametes, that fusion is isogamous or anisogamous and that reduction division occurs when the zygote germinates.

Vaucheria is a sparingly branched tubular thallus with a central tubular vacuole surrounded by a peripheral layer of cytoplasm, and external to that cell wall. The lining layer of cytoplasm contains a large number of small nuclei together with numerous small circular or elliptical chloroplasts without pyrenoids.

The taxonomic position of this genus has been in dispute for many years, and until 1945 it was included in the Caulerpales of the Chlorophyceae. There are three main reasons for excluding it from the Chlorophyceae. 1. The carotenoids tend to be in excess of the chlorophylls and none of the carotenoids characteristic of the Caulerpales is present. Also chlorophyll *b* is absent and recently chlorophyll *e* has been identified in the zoospores[224] (but not in any other part of the plant). 2. Starch never accumulates. 3. The flagella are unequal and the shorter one of the spermatozoid is of the pantonematic type (the zoospores does not possess a pantonematic flagellum). Although its inclusion in the Xanthophyceae appears well established, its elaborate oogamous sexual reproduction is unlike any process in the remainder of the Xanthophyceae, so that it is difficult to establish any affinity between *Vaucheria* and other genera in the class.

Asexual reproduction most commonly occurs by the production of large multiflagellate zoospores. A sporangium is formed by the swelling of the distal end of a branch followed by the development of a transverse wall at the base of the swelling. The protoplast (containing numerous nuclei and chloroplasts) contracts away from the sporangial wall, and 2 flagella develop opposite each nucleus. The mature zoospore is then extruded through a pore at the distal end of the sporangium, swims for a short time, comes to rest and, after losing its flagella, germinates by the development of a number of outgrowths which may elongate considerably. Alternative methods of asexual reproduction such as the formation of aplanospores, thin-walled akinetes or thick-walled hypnospores can be induced by specific environmental conditions.

All species reproduce sexually; some are heterothallic and others homothallic. In the latter species the antheridia and oogonia are frequently borne on the same branch, or sometimes on adjacent branches. Both of the sex organs begin as a swelling on the side of a branch; the oogonium remains as an ovoid swelling, but the antheridium usually grows to form a hooked tube. The distal end of the tube becomes cut off by the formation of a transverse wall, and within this distal portion (containing many nuclei and a few chloroplasts) the protoplast divides to form a large number of uninucleate, biflagellate spermatozoids. The oogonium, on the other hand, contains a single, large, uninucleate ovum. The uninucleate situation probably arises from the migration of all nuclei (except one) back into the main part of the thallus before formation of the cross wall.

The spermatozoids swim to the oogonium and enter it through an apical pore formed by gelatinization of part of the oogonial wall. Only one spermatozoid enters the egg, the male nucleus migrates to the female, and, after increasing in size, fuses with it. The zygote wall usually thickens and there is usually a resting period before germination.

7

CHRYSOPHYTA:

CHRYSOPHYCEAE AND HAPTOPHYCEAE

Division CHRYSOPHYTA

Classes CHRYSOPHYCEAE and HAPTOPHYCEAE

Species of the Chrysophyta are characterised by four main features. 1. The photosynthetic pigments include chlorophyll *a* and carotenoids such as fucoxanthin and diadinoxanthin. 2. The flagellation is variable; the cells may be either uniflagellate or biflagellate, and when biflagellate, they may be either equal or unequal in length and similar or dissimilar in nature. 3. The main reserve products are oil and chrysolaminarin (leucosin); starch never accumulates. 4. Several featyres of cell structure are characteristic; for example, a cellulose wall is absent and the membrance shows a marked tendency to become silicified, also the production of endogenous cysts is common.

Numerically, species of this division constitute some of the most abundant kind of life on earth, being common components of phytoplankton from a number of different habitats (chiefly freshwater).

Although the Chrysophyta consists mainly of unicellular flagellate forms, the presence of vegetative forms similar to many found in the Chlorophyta and Xanthophyta is of interest as a possible indicator of parallel lines of evolution, and most authorities emphasise this aspect of Chrysophyta. More recent investigations with the electron microscope have emphasised details of cell ultrastructure, and these indicate that there are two series of organisms in the Chrysophyta. Christensen[26] has given each of these the status of a class; the one being called the *Chrysophyceae* and the other the *Haptophyceae*. At the present time it is not possible to discuss each class separately, since many species of Chrysophyta have not been examined with the electron microscope, and cannot be assigned to either of the classes with any certainty.

Thus, in the present discussion of the Chrysophyta, two aspects are emphasised. First, details of cell structure are presented with particular reference to those used to separate Chrysophyceae from Haptophyceae. Secondly, developments of vegetative form paralleling those in the Chlorophyta and Xanthophyta are also described.

CELL STRUCTURE AND THE DISTINCTION BETWEEN CHRYSOPHYCEAE AND HAPTOPHYCEAE

The classification of the Chrysophyta into two classes is based on two main characters: first, the nature of the flagella; second, the nature of the scales on the cell surface.

Flagella

Cells of the Chrysophyta show three kinds of flagellar arrangement: 1. uniflagellate, with the flagellum of the pantonematic type (fig. 16 A); 2. biflagellate with a pantonematic flagellum, which is longer than the other of the acronematic type (fig. 16 B); 3. biflagellate with both flagella of the acronematic type, and a third appendage (*haptonema*) also present in some species (fig. 16 C-E).

The distinction between the third kind of arrangement and the others appears to be well defined, whereas the distinction between conditions (1) and (2) is not so.[150] Thus, Bourrelly[14] suggests that within the series possessing a pantonematic flagellum there is a progressive reduction in the size of the acronematic type until in *Mallomonas* it exists solely as the photoreceptor peduncle and in *Chromulina* it has disappeared completely. This suggestion is supported by the observation that *Chromulina psammobia* has a short internal second flagellum.[78, 207] Hence, on the basis of flagella form the Chrysophyta can be divided into two series, Chrysophyceae and Haptophyceae, the former possessing a pantonematic flagellum with or without a second acronematic one, and the latter possessing two acronematic flagella.

Some species of the Haptophyceae possess a third filiform appendage apparently used for the temporary anchorage of the cell. This appendage has been termed a *haptonema*. It does not appear to have a locomotory function and may be extremely long

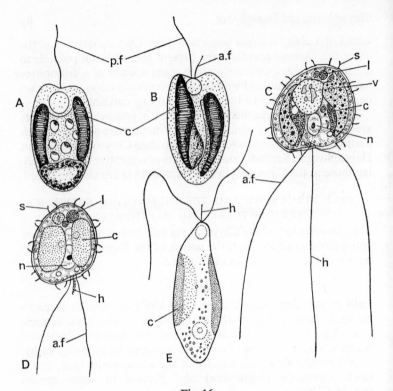

Fig. 16
Chrysophyta: cell structure of Chrysophyceae and Haptophyceae

A, *Chromulina ovalis*. B, *Ochromonas mutabilis*. C, *Chrusochromulina chiton*, swimming with haptonema extended. D, *C. chiton*, stationary with haptonema tightly coiled. E, *Prymnesium saltans* showing short haptonema. a.f, acronematic flagellum; c, chloroplast; h, haptomena; l, leucosin vesicle; n, nucleus; p.f, pantonematic flagellum; s, scale; v, vacuole. (A, B after Fritsch[90]; C, D after Parke *et al.*[194]; E after Fott[84])

(e.g. in *Chrysochromulina*, where its length varies from 2 to 20 times the body length, fig. 16 C). In some species (e.g. *Chrysochromulina chiton*, fig. 16 D) it is coiled when not anchoring the cell, whereas in *Prymnesium* (fig. 16 E) it is permanently extended. The detailed structure of the haptonema is different from that of a flagellum, since in cross-section it does not show the '9+2' arrangement of component fibrils, but is seen to consist of three concentric membranes surrounding a ring of fibres and a central space. It is not known how general the presence of a haptonema might be, but one cannot eliminate the possibility that it is present in all species, at least in a rudimentary form not recognisable in the light microscope.

Examples of genera in the Haptophyceae include *Isochrysis, Deacronema, Dicrateria, Prymnesium, Phaeocystis, Chrysochromulina*; the Chrysophyceae contain such genera as *Ochromonas, Synura, Paraphysomonas, Mallomonas* and *Chromulina* (more detailed lists can be found in ref. 191).

Structure of the scales

Both the Chrysophyceae and Haptophyceae have 'naked' species in which the cells are bounded only by the cytoplasmic membrane, whereas in other species the cell surface is covered by a layer of delicate scales. These scales must not be confused with the much larger coccoliths of the Coccolithophoridaceae (see below) since they cannot be seen with the light microscope and are only visible when examined with the electron microscope. A series of observations with the electron microscope (particularly by Manton, Parke and their colleagues[153, 160, 161, 162, 163, 193, 194, 195]) has shown the scales to have an elaborate structure, and has further shown important differences in scale structure between the Chrysophyceae and the Haptophyceae. In the former, the scales are impregnated with silica, and their structure is variable and shows no basic type of construction. The delicate scales of the Haptophyceae do not contain silica but appear to consist solely of organic material. Moreover the basic scale structure is uniform and different from that found in the Chrysophyceae; this basic type consists of a round or oval scale, the outer face of which has a lattice of cross striations within a raised rim and the inner face has a characteristic pattern of radiating ridges. Usually, only one layer of scales is present, although many species of *Chrisochromulina* (Haptophyceae) possess a second layer of cup-shaped scales arranged as a layer outside the layer of plate-like scales.

The origin of the scales remained a mystery for many years after their discovery but several pieces of evidence[162, 166] now indicate that they originate at the Golgi apparatus. However, the details of their formation and their passage to the surface of the cell are not clear.

Place of the Coccolithophoridaceae

A large group of marine species of Chrysophyta possess plates of calcium carbonate (*coccoliths*) embedded in the gelatinous envelope. These *coccolithophorids* are abundant in the sea and their deposition on the sea bottom has been responsible for the formation of the extensive chalk beds. When the Chrysophyta was first divided into two distinct series the Coccolithophorids were not included, since no evidence was available to classify them. However, increasing evidence supports the claim that they belong to the Haptophyceae, and Parke[189] has summarised this evidence as follows:

1. Included among the original species described having two acronematic flagella was a coccolithophorid, *Cricosphaera carterae*.

2. von Stosch[231, 232] observed that a swarmer produced by a *Syracosphaera* type of coccolithophorid was similar to *Prymnesium*.

3. Parke and Adams[190] observed that a motile phase of *Coccolithus pelagicus* has two acronematic flagella, a haptonema of the *Chrysochromulina* type and scales with a radiating pattern of ridges of the inner face.

4. Motile phases of three other Coccolithophorids, *Cricosphaera carterae*, *Cricosphaera sp.* and *Pleurochrysis scherffelii*, also have affinities with the Haptophyceae, since they possess two acronematic flagella and a short haptonema.

Other features of cell structure

The above remarks have emphasised those features of the cell structure used to separate Chrysophyceae from Haptophyceae and has omitted any reference to the many other interesting features. Furthermore such features have sometimes been neglected in the recent observations with the electron microscope so that many of the remarks made in this section refer to observations with the light microscope and confirmation and elaboration of these observations with electron microscopic examination is desirable. Without such an examination it is not possible to assign a particular species to the Chrysophyceae or the Haptophyceae, so most of the remarks in the remainder of the chapter apply to the entire division, Chrysophyta. Further observations may show important differences between the two classes with regard to many other features.

Cell surface

Most motile forms are naked and the protoplast is bounded by a *periplast*. The periplast is normally soft and permits considerable changes of shape of the cell, although in species such as *Chromulina verrucosa* and *Chromulina pyrum* the periplast is firm and the cells have a fixed shape. In the coccoid forms (Chrysosphaerales) the cells have a well-defined wall, although its composition is unknown.

Among the motile forms there are three main modifications to the simple periplast:

1. In the *Coccolithophoridaceae* the surface layers are gelatinous at first but later many calcified discs (*coccoliths*) are deposited over the entire surface (fig. 17 A). Coccoliths are of two kinds, the *heterococcoliths* in which the crystallite fibrils are arranged radially on a disc, and the *holococcoliths* consisting of the rhombohedral or hexagonal crystal type. The mechanism of deposition of the calcium carbonate, as in all calcareous algae, is unknown and the various hypotheses and available evidence have been summarised by Lewin.[145] It is commonly assumed that removal of carbon dioxide from the surrounding water by photosynthesis alters the ionic balance and calcium carbonate is deposited at the cell surface. However, if this were the sole mechanism, it is difficult to understand why *all* marine algae do not have calcareous deposits. The site of origin of the coccoliths is also unknown, although evidence is accumulating that they are formed inside the cell rather than at the cell surface. Thus, Parke and Adams[190] obtained evidence suggesting that coccoliths originated in vesicles deep within the cell, although the method of formation and the subsequent arrangement on the cell surface are unknown. Wilbur and Watabe[242] have also observed that coccoliths are formed one at a time in a clearly defined area of the cytoplasm termed the 'matrix'. The observations of Manton and Leedale,[164] however, do not support this idea. These workers found that the minute scales on the surface of *Crystallolithus hyalinus* originated in internal vesicles, but there was no evidence that the holococcoliths originated in the same way. Instead the authors suggested that 'they may be developed outside the cell from calcite secreted in solution into the appropriate cavity'.

The process of coccolith formation appears to be controlled by the cytoplasm since in heterococcoliths the shape of individual crystals is completely subordinated to the architecture of the coccolith as a whole.

The rates of photosynthesis and coccolith production show the same relationship to light intensity. From this it was concluded that the two processes were intimately linked, and a working hypothesis was proposed that carbon was taken up by the cells as bicarbonate and that molecular carbon dioxide was extracted from the bicarbonate during photosynthesis so that carbonate remained and was precipitated as a coccolith.[183, 227] Extensive investigations of Paasche[184] did not support this hypothesis and suggested instead that the accumulation of calcium carbonate by coccolith-forming cells depended, at least in part, on the supply of adenosine triphosphate (ATP) by cyclic photophosphorylation.

2. A second elaboration of the periplast is found in members of the Mallomonadaceae. The surface layers are distinguishable as two distinct layers, the inner of which appears to be cellulose and the outer one of a pectic nature. In *Microglena* numerous lens-shaped scales of silica are embedded in the pectic layer and in *Mallomonas* the structure is further elaborated by each scale bearing a long silicified needle (fig. 17 B). The possession of such needles is assumed to be an adaptation to the planktonic existence of *Mallomonas*.

3. A third elaboration of the cell surface is found in the *loricate*, or *encapsuled*, forms. In these the protoplast is surrounded by a rigid cell envelope and separated from it by a space (fig. 17 G). It appears likely that such loricate forms are present in both the Chrysophyceae and the Haptophyceae.

Chloroplasts

The cell contains parietal chloroplasts and normally they are few in number, usually being one or two. Frequently it is difficult to distinguish between a single two-lobed chloroplast and two chloroplasts. For example, the chloroplasts of *Mallomonas ouradion*, *Chrysosphaerella brevispina* and *Synura sphagnicola* were originally described as being double but were later shown to be single with a narrow connection between two lobes.[109]

The form of the chloroplast in *Chrysochromulina* depends on the stage in the life-history; motile stages contain saucer-shaped parietal chloroplasts, whereas the non-motile stages have stellate chloroplasts.

Pyrenoids may be present (e.g. *Chromulina*, *Mallomonas*) or absent (e.g. *Ochromonas*), and they sometimes show variability in a single genus. *Chrysochromulina parva* has a pyrenoid within the chloroplast whereas that of *C. chiton* is outside the chloroplast. Since starch is not deposited in species of Chrysophyta, it is not known whether the pyrenoid is functionally equivalent to that in the Chlorophyta.

An electron microscopic examination of the chloroplast of *Ochromonas danica*[96] has shown it to consist of a granular matrix traversed by bands, each of which consists of three discs. Thus the structure is similar to that in most algal divisions and does not show the formation of grana as in the chloroplasts of higher plants.

When cells of *Ochromonas* are grown in the dark with organic substrates the cells lose their photosynthetic pigments and electron microscopic examination confirms that this loss of pigments is accompanied by a loss of the chloroplasts. That is, no leucoplasts are present in the colourless cells as are present in the cells of *Polytoma* (p. 42) and the loss of chloroplasts resembles that found when *Euglena* is grown under heterotrophic conditions (see p. 123). Many species of the Chrysophyta lose their chloroplasts permanently and colourless species of *Chromulina* and *Ochromonas*, etc., are common. The cells of many of these species show amoeboid changes of shape and a holozoic mode of nutrition is common amongst such forms.

Contractile vacuoles and stigma

In common with most motile unicellular algae, such forms in the Chrysophyta also possess an eye-spot (stigma) and contractile vacuoles. The presence of an eye-spot does not appear to be widespread and no details of its structure and function are available. Contractile vacuoles are of variable form; there are two simple vacuoles located at the anterior of the cell in *Chromulina*, whereas in *Microglena* the vacuolar apparatus is more complex, consisting of a large anterior non-contractile reservoir opening to the exterior by a narrow canal and a number of smaller contractile vacuoles arranged around and connected to this central reservoir. In the genus *Chrysochromulina* contractile vacuoles are found in freshwater species (*C. parva*) but not in the marine species.[193]

Nucleus

Cells of the Chrysophyta are uninucleate but details of the structure and mode of division have not been studied extensively. The nucleus has a well-defined membrane and a prominent nucleolus. A connection between the flagella bases and the nucleus has been observed in *Paraphysomonas vestita*,[162] but it is difficult to decide on the possible significance of this connection. Manton and Leedale[162] suggest a mechanical function since fibres also connect the nucleus to the cell wall and in other algae the flagella bases are attached to the chloroplast and to the cell walls. Similar fibrous connections have been reported between the flagella bases and the nucleus in other phyla, but their significance is unknown.

Reproduction of the motile unicellular species

The most common method of reproduction is by longitudinal division of the cell. This can occur when the cell is motile, but more commonly the cell loses its flagella before the onset of division. In some species (e.g. *Chromulina rosanoffii*) the daughter cells fail to separate and extensive palmelloid masses can be formed.

A characteristic feature of the Chrysophyta is the presence of *cysts* (statospores). At the onset of cyst formation the flagella are retracted, the cell becomes spherical and the protoplast contracts away from the cell envelope. A new gelatinous envelope is formed around the protoplast and later a cellulose wall is produced. This wall later becomes thickened and impregnated with silica except for a small pore which becomes filled with a gelatinous plug. This thick-walled cyst appears to constitute a resting stage. At germination the amoeboid protoplast is liberated through the pore, flagella are formed and the normal vegetative stage is thus produced. In some species the contents of the cyst divide to liberate 2–4 zoospores.

Many early records of sexual reproduction were based largely on two kinds of observations. 1. The observation of two cells with their protoplasts apparently joined. 2. The frequent observation of 2 nuclei in a cyst. Although both have been assumed to indicate fusion of cells, both observations can also be interpreted as being stages in the asexual or vegetative method of reproduction. It appears, however, that an isogamous method of reproduction (variable in its details) is generally accepted for the Chrysophyta (the detailed arguments for this interpretation of paired cells of motile loricate species is given by Lund[149]). Matviyenko[169] describes three kinds of isogamous process in the Chrysophyta:

1. *'Typical' Isogamy*. The production of motile gametes which fuse as in the isogamous process found in, for example, the Chlorophyceae. This method occurs in *Ochrosphaera neapolitana*.

2. *Hologamy*. The fusion of the protoplasts of two vegetative cells. This is the method discussed by Lund for such loricate genera as *Kephyrion* and *Calycomonas*.

3. *Autogamy*. In this process there appears to be fusion of two

nuclei within the cyst but it does not appear to have been preceded by fusion of two protoplasts.

In all accounts of the sexual process the zygote is described as a typical cyst-like structure. Division of the nucleus at the germination of the cyst-like zygote appears to be a reduction division and 4 daughter cells are normally produced.

RANGE OF VEGETATIVE FORM

Although motile unicells constitute the bulk of the Chrysophyta the presence of 'algal' types has been considered an important feature since it permits the possible evolutionary developments in this group to be compared with those in other classes, notably the Chlorophyceae and Xanthophyceae. In addition to the motile unicellular forms one can identify the development of the *motile colonial form*, the *coccoid*, *palmelloid* and *dendroid* forms, and also the filamentous habit. Most authorities classify the Chrysophyta into a number of orders based on the vegetative structure; *Chromulinales* and *Ochromonadales* (unicellular and colonial motile forms), *Chrysosphaerales* (coccoid), *Chrysocapsales* (palmelloid), *Rhizochrysidales* (rhizopodial and dendroid) and the *Phaeothamniales* (filamentous).

Although the range of vegetative forms is to be discussed in this present chapter, the taxonomic implications are not considered. There appear to be two reasons for not proposing a rigid taxonomic scheme. First, work on the life-histories of some coccolithophorids (described at the end of this chapter) casts in doubt many of the basic assumptions of a classification system based on vegetative structure. Second, any scheme must be altered when it becomes possible to assign more species to one or other of the two classes, the Chrysophyceae or the Haptophyceae. Such a new classification of the two groups separately has already been proposed.[191]

The form of the *motile colonies* is extremely variable. For example, *Chrysosphaerella* is a spherical colony with cells grouped in a mucilaginous envelope in which silicified scales are embedded; long hollow needles radiate from the colony, each one arising from a cell near the base of the flagellum. In *Synura uvella*, however, there is no enveloping mucilage and the ellipsoidal cells are united by their posterior parts by the localised production of mucilage (fig. 17 C). None of the colonies show the degree of complexity found in the Volvocales. Thus, they possess no distinct anterior end, and asexual reproduction occurs by fission of the colony.

Genera showing the *coccoid* tendency include *Chrysosphaera* and *Epichrysis*, in both of which the cells are large and spherical and, in common with most non-motile forms of the Chrysophyta, possess a distinct cell wall. The cells sometimes occur singly although they also frequently occur in groups of varying sizes.

The *rhizopodial* form is an extension of the amoeboid tendency shown by many motile unicellular species of the Chrysophyta. In *Chrysamoeba*, for example, the cells can be flagellated and so

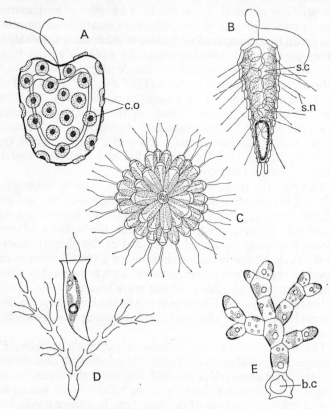

Fig. 17 Range of morphological types in the Chrysophyta

A, *Hymenomonas roseola*. B, *Mallomonas fastigiata*. C, *Synura uvella*. D, *Dinobryon pediforme*. E, *Phaeothamnion convervicola*. co, coccolith; s.c, silicified scale; s.n, silicified needle. (A after Fritsch[90]; B–E after Fott[84])

resemble *Chromulina*; but sometimes the flagella are lost, the cell becomes completely amoeboid and a number of rhizopodia are put out. In many species these rhizopodia anchor the cells to the substratum. In other species (e.g. *Chrysidiastrum*) cells with delicate radiating rhizopodia are united by coarse protoplasmic strands to form short chains of cells.

The *dendroid* colony is one in which cells are attached to each other by prolongations from various parts (usually the posterior) of the cell surface and attached to the substratum by similar stalks. The simple development of such a form is found in many species of *Ochromonas* which attach themselves by a pseudo-podium. However, the stage is transient, whereas in *Chrysodendron* the cells are anchored to the substratum permanently and a number of cells are attached together by delicate elongated stalks. *Dinobryon* is a common planktonic genus showing a dendroid type of colony in which the cells are surrounded by a wide envelope pointed at the base and open at the top (fig. 17 D). Unlike the rhizopodial forms, the cells of all dendroid colonies retain their flagella and the colonies frequently break up and the individual motile cells swim away.

The *palmelloid* type of structure is highly developed in the Chrysophyta and the plant body can reach a high degree of elaboration. *Chrysocapsa* is a relatively simple form in which spherical cells are embedded in a structureless spherical or ellipsoidal mass of mucilage. In *Hydrurus*, however, the numerous naked cells are embedded in long cylindrical strands of mucilage and elaborate feathery tufts are formed. Growth is apical, with a single apical cell dividing transversely, half becoming the new apical cell and half elongating to form a new vegetative cell. Both *Chrysocapsa* and *Hydrurus* belong to the Chrysophyceae whereas another palmelloid genus, *Phaeocystis*, a planktonic genus consisting of spherical cells embedded in spherical, lobed, mucilaginous masses, is a member of the Haptophyceae.

Filamentous forms include unbranched chains of cells as in *Nematochrysis*, and the short branched (or unbranched) threads of *Phaeothamnion* (fig. 17 E). Another genus, *Phaeodermatium*, is discoid and Fritsch[90] speculates that this might represent the prostrate system of a heterotrichous species.

The various non-motile forms described above are generally identified as members of the Chrysophyta mainly on the basis of the appearance of the motile stage, since they all reproduce asexually by the production of zoospores. Characteristic cysts are

D

also formed, although their relationship to any possible sexual process is unknown.

Life-history of some coccolithophorids

Parke[189] has described changes of form when three species of Coccolithophoridaceae, *Cricosphaera carterae*, *Cricosphaera sp.* and *Pleurochrysis scherffelli*, are maintained in laboratory culture. Since the observations may have far-reaching effects on the understanding and classification of the Chrysophyta, they are presented in some detail.

The 'naked' motile forms are a convenient starting point. At first they show the Haptophycean characters mentioned in the beginning of the chapter. Later they acquire their coccoliths, and these motile coccolith-containing stages are identified as the species mentioned above, and they may be maintained as such by continuous sub-culturing. However, when sub-culturing is stopped, various non-motile stages can be observed and it is possible to compare these forms with descriptions and illustrations of type specimens of a variety of genera. For example, when the flagella are first lost the cells resemble the non-motile coccolithophorid, *Ochrosphaera*. Later the cell contents divide to give four daughter cells in a tetrad, and after the coccoliths are discarded, forms apparently identical to the coccoid *Chrysosphaera* can be observed. When the cells of the tetrad slide past each other to form short chains, they divide and show strong resemblances to the filamentous genera *Nematochrysis* or *Thallochrysis*. Later, after further division, filamentous forms apparently identical to descriptions of *Apistonema* and *Chrysonema* can be observed. Finally, periodic contractions of the protoplast in an upward direction followed by secretion of the new membrane produce dendroid colonies of the *Chrysotila* type.

Thus, within the life-history of a single species one can recognise stages showing almost the entire range of vegetative structure found in the Chrysophyta. Since the vegetative form is the main criterion for the classification of the Chrysophyta into its orders, these observations may have considerable effect on the taxonomy of the Chrysophyta. They afford another striking example of the need for detailed observation of many algae in laboratory culture before a real understanding of their various interrelationships can be obtained.

8

BACILLARIOPHYTA: BACILLARIOPHYCEAE

Division BACILLARIOPHYTA

Class BACILLARIOPHYCEAE (Diatoms)

Diatoms are extremely widespread and occur as the dominant organisms of many diverse habitats. They are particularly conspicuous in both marine and freshwater phytoplankton populations, and are prominent in soil and the bottom flora of lakes and ponds. They are characterised by four main features: 1. The cell walls are silicified and show characteristic secondary structures (see below). 2. The photosynthetic pigments include chlorophylls *a* and *c,* together with the xanthophyll, fucoxanthin. 3. Food storage products include fats and chrysolaminarin. 4. The motile states possess a single pantonematic flagellum.

Although it has been possible to identify a number of structural developments in the Xanthophyta and Chrysophyta paralleling those in the Chlorophyta, the range of form has been much more restricted. The reduction in the range of vegetative types is even more marked in the diatoms, in which only unicellular and colonial forms can be recognised.

STRUCTURE OF THE DIATOM CELL

The diatom cell wall (frustule) consists of two parts, the one fitting over the other as does the lid over the main part of the box (fig. 18 A). The outer part (the 'lid') is the *epitheca* and the inner part is the *hypotheca.* Each half consists of the main surface, the *valve,* and the overlapping *connecting bands*; the two bands constitute the *girdle*. The diatom cell can therefore be viewed from two directions, the girdle view and the valve view (fig. 18 B, C). The axis between the middle of the two valves is the long axis (most diatoms are broader than long) and is called the *pervalvar* axis, and the one at right angles to this is the *valvar* plane.

All diatom frustules have silicified walls, although in *Phaeodactylum tricornutum* only one valve is silicified.[147] The walls contain

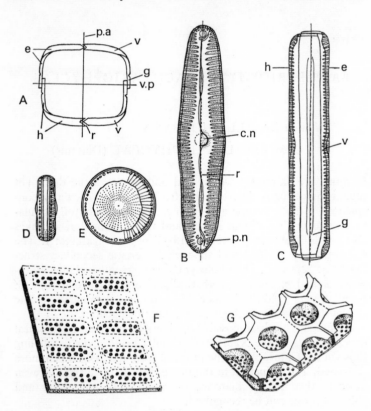

Fig. 18 Bacillariophyceae

A, B, C, *Pinnularia viridis*: A, transverse section; B, valve-view; C, girdle-view. D, E, *Cyclotella comta*: D, girdle view; E, valve view. F, laminar wall. G, locular wall. c.n, central nodule; e, epitheca; g, girdle; h, hypotheca; p.a, pervalvar axis; p.n, polar nodule; r, raphe; v, valve; v.p, valvar plane. (A–E after Fritsch[90]; F, G after Hendey[114])

hydrated silica and, since no other element can replace silicon, growth of diatoms show an absolute requirement for silicon, and if other elements are present in adequate amounts, growth is proportional to the silicon concentration. In the more highly silicified forms, silica can represent about 50 per cent of the dry weight of the cells,[196] whereas in *Navicula pellicosa* it may be as low as 4 per cent[143] and in *Phaeodactylum tricornutum* less than 1 per cent.[147] The silicified wall also contains an organic component, which has been called 'pectin', although there is no definitive chemical evidence for this statement. The process of silicification is presumably controlled by the cytoplasm, and the accumulation of silica is an energy—requiring process.[144] There are no data on the mechanism of its deposition. It appears to be extremely rapid since new cell walls may be deposited and completely silicified within 10–20 minutes of the division of the protoplast.[144]

Because of the frequent high degree of silicification the problem of buoyancy of planktonic diatoms has been discussed by many authorities (e.g. ref. 148). Three main theories have been proposed, although none has overwhelming experimental support. One hypothesis suggests that accumulation of oil and fats in the cytoplasm is important. However, under conditions of nutrient deficiency, when the proportion of fats tends to increase, the rate of sinking of *Skeletonema costatum* increases.[226] A second hypothesis emphasizes the importance of morphological modifications which tend to change the shape of the cell from spherical; for example, the production of spines, the formation of colonies, etc. A third suggestion, made by Gross and Zeuthen,[105] is that large marine planktonic diatoms decrease their specific gravity by excluding metal ions from their cell sap. However, Lund[148] has pointed out that it is difficult to apply it to freshwater forms, since the ion content of fresh waters is so low that no appreciable gain in buoyancy would be achieved by excluding ions.

The diatom cell wall is not of uniform thickness, and periodic arrangements of thicker and thinner areas produce a complex series of 'markings' on the cell surface. The general arrangement and symmetry of these 'markings' are important criteria for the division of the Bacillariophyceae into the two orders, the *Centrales* and the *Pennales*. In the former they are arranged with reference to a central point (fig. 18 E) (this basic picture may be altered when the cell is angular). In the latter, however, the main structural element on the cell surface is a spine, and the finer secondary structures are arranged as lateral branches (fig. 18 B). Thus, the valve of pennate forms has a narrow *axial area* thickened at each end (*polar nodule*), and a dilated *central area* usually having a median thickening (*central nodule*). The axial region sometimes has a slit (*raphe*) which runs from one polar nodule to the other; whereas in other species a lighter region of the axial area gives the superficial appearance of a raphe, and this is called a *pseudoraphe*.

The secondary structures of the diatom cell wall represent the fine sculpturing which occurs over much of the valve surface. These structures are extremely variable; their nomenclature is sometimes equally variable and much confusion has resulted. Hendy[114, 115] has attempted to standardise the terminology, and his system will be used as a basis for the present discussion. Examination in the electron microscope has confirmed earlier observations of the light microscopists in showing four basic kinds of secondary structures: 1. *punctae* are small perforations of the valve surface and are frequently arranged in regular lines, or *striae*, 2. *aerolae* are larger, depressed box-like structures, 3. *canaliculi* are narrow tubular channels through the valve wall and 4. *costae* are ribs formed by the heavy deposition of silica.

From the voluminous literature on the detailed arrangements and elaborations of these four basic kinds of structures, Hendy has recognised two main wall types: first, the *laminar* walls consisting of a single silicified layer with a great variety of patterns on it (fig. 18 F), and second, the *locular* walls consisting of two parallel layers with a complex reticulum of cross walls between them (fig. 18 G). Examples of species with lamina walls include *Synedra fulgens*, *Fragilaria construens* and *Didymosphenia geminata*. Species in which the wall is of the locular type include *Triceratium favus*, *Coscinodiscus asteromphalus* (in both of these species the locular nature of the wall can be seen with the light microscope) and *Stephanopyxis palmeriana*.

The general appearance of the diatom cell wall is one of openness, since the total area of the holes may vary from 10–30 per cent of the total valve surface.

OTHER ASPECTS OF CELL STRUCTURE

The above discussion has been concerned solely with the structure of the wall, and in fact most observations with the electron microscope have emphasised the wall structure of diatoms. Because of this, the structure of the cytoplasm has been largely neglected in recent investigations. Many of the structures outlined below have therefore been summarised from older work with the light microscope and confirmation of many of the features by electron microscopic examination is desirable.

The cytoplasm normally occurs as a lining layer surrounding a large central vacuole. The cells are uninucleate and in the Pennales the nucleus is normally suspended in the vacuole within strands of cytoplasm. In the Centrales, however, the nucleus most commonly occurs in the lining layer of cytoplasm. Division of the nucleus is mitotic but the process has two interesting features: first, an intranuclear spindle is present at metaphase and the chromosomes are arranged as an equatorial ring rather than as a plate[94] (a ring arrangement appears to be more common in animals than in plants); second, the spindle is in the form of an elongated cylinder with flattened poles,[94] and this compares with a similar structure in the nuclei of Chrysophyta and Pyrrophyta.[19]

Chloroplasts are usually parietal. In the Centrales a cell usually contains numerous chloroplasts which frequently (e.g. *Biddulphia*) appear as minute granules, although in *Melosira* they appear as large lobed discs. The cells of Pennales normally have a different form of chloroplast, namely a large plate-like chloroplast which may sometimes be lobed. Electron microscopic examination of the chloroplasts[238] has shown them to consist of a matrix traversed by a number of bands, each of which consists of 4–6 lamellae. As we have seen before (p. 41), this type of structure is general throughout most divisions of algae and contrasts with the grana structure of higher plant chloroplasts.

Other subcellular structures such as mitochondria, oil droplets, pyrenoids and Golgi bodies have also been observed.

GROSS VEGETATIVE STRUCTURE

The general shape and superficial appearance of the diatom cell is variable and many of these gross variations are useful for a preliminary identification of a given alga. Fritsch[90] outlines six main types of morphological elaborations which appear to be correlated with the planktonic habit: 1. the flat discoid shape of many centric diatoms (e.g. *Cyclotella comta*, fig. 18 D, E); 2. the needle shape of species such as *Rhizosolenia* and *Synedra*; 3. the long, sometimes coiled filaments of some species of *Melosira*; 4. the development of elongated bristles (*Stephanodiscus*) or horns (*Chaetoceros*) from the edge of the valves; 5. the stellate colonies of *Asterionella* and *Tabellaria*; and 6. the frequent production of extensive mucilaginous envelopes (e.g. *Cyclotella planctonica*).

Although most diatoms are unicellular, a number of colonial species is also known. There appear to be three basic kinds of colony formation. Firstly, cells are sometimes joined together by a localised production of mucilage to form stellate colonies of *Asterionella* of the filamentous forms of *Melosira*. Secondly, some colonies consist of many cells embedded in a common mucilaginous envelope, and in some genera the envelope has a tubular structure (e.g. some species of *Navicula*, *Cymbella* and *Nitzschia*). The third type of colony consists of cells joined by special outgrowth such as the spines of *Chaetoceros*.

CELL DIVISION

Diatoms usually divide at night and the plane of division is always at right angles to the longitudinal axis, that is, parallel to the valve surfaces. The first indication that cell division is imminent is that the cell increases in size and the two halves separate slightly. Mitotic division of the nucleus is followed by fission of the protoplast in a plane parallel to the valve faces. New siliceous valves are then deposited on the two fresh protoplasmic surfaces. As the connecting bands develop, the valves of the parent cell separate and the new silica valve becomes the hypotheca of each daughter cell.

The daughter cell having the original hypotheca of the parent (it is now the epitheca of the daughter cell) is smaller than the parent cell. Thus, in a population of diatoms there is normally a progressive decrease in the average cell size (those species which do not show a progressive decrease in size are usually weakly silicified and the constancy of size is probably related to the plasticity of the cell wall).[116]

AUXOSPORE FORMATION AND SEXUAL REPRODUCTION

The continued decrease in cell size in a population of diatoms is prevented by the formation of *auxospores*. Auxospores are large cells which develop from newly formed zygotes (before the process of sexual reproduction of centric diatoms became clarified auxospore formation in the Centrales was thought not to depend on the fusion of gametes). There appear to be no confirmed reports of asexual auxospore formation and their formation is intimately related to the process of sexual reproduction.[146] The process of sexual reproduction is therefore linked to reduction in cell size and, for a given species, there appears to be a distinct size range (usually 30–40 per cent of maximum) within which the cells are capable of becoming sexual. In those diatoms which do not decrease in size sexual reproduction does not apparently occur.[41]

The vegetative cells of both Pennales and Centrales are diploid and gametes are formed directly following meiosis. However, the sexual process shows a striking difference between the centric and the pennate diatoms (see reviews of Geitler,[93] Patrick,[197] Grell,[104] Lewin and Guillard).[146] In the Pennales the process is isogamous and the gametes are amoeboid whereas in the Centrales the process is oogamous and the spermatozoids are flagellated.

In the pennate forms, cells aggregate in pairs (*gamontogamy*), nuclear division (meiotic) produces 4 nuclei, and of which only 1 or 2 normally develop into gametes. The parent cells fuse (*cystogamy*) and the fusion of gametes (generally derived from different cells) occurs in a copulatory jelly. In most species, fusion is isogomous and the gametes move towards each other. However, some species show physiological anisogamy, in which, although the gametes are morphologically identical, one is stationary whereas the other is motile.

In the Centrales, oogonia occur as slightly extended cells with an elongated nucleus. Normally, each oogonium contains only one egg nucleus, since one nucleus degenerates at the end of each of the two divisions. The egg cell normally remains enclosed in the oogonium, or attached to it, until fertilisation. Four uniflagellate spermatozoids are formed by division of the contents of the male sex organ, the spermatogonium. However, the formation of the spermatogonium itself is variable; in *Cyclotella tenuistriata*, for example, the vegetative cell acts directly as a spermatogonium, whereas in other species (e.g. *Melosira varians*) the vegetative cell divides to form 4–8 spermatogonia.

In both pennate and centric diatoms the zygote begins to grow immediately after fusion and develops into an auxospore. Two mitoses, at each of which only one daughter nucleus survives, ultimately give rise to the vegetative cell which has the largest dimensions of the particular species.

There are two known exceptions to the generalised scheme of sexual reproduction outlined above. First, *Rhabdonema adriaticum* is a pennate diatom which is dioecious (all other diatoms, both pennate and centric, appear to be monoecious), and the sexual process is oogamous.[233] Female cells produce a single egg per oogonium, and male cells produce spermatogonia which are dispersed passively through water, before becoming attached by mucilage pads to the oogonium. They then produce naked amoeboid non-flagellate microgametes. At fertilisation only the nucleus is injected into the egg cell. A second exception is found in the Centrales, where gametes have not been observed in *Cyclotella meneghiniana* and in *Melosira nummuloides*. However, both species produce auxospores and they might represent isolated examples of asexual auxospore formation; alternatively, it has been suggested[146] that intracellular autogamy might occur in these species.

CLASSIFICATION OF DIATOMS

As pointed out in the beginning of the chapter the diatoms can be divided into two orders, the Centrales and the Pennales; this classification being based largely on the symmetry and orientation of the secondary structures on the valve surface. Although Hendey[112, 114, 115] considers this distinction to be artificial, it appears to be well founded since three other features parallel the differences in cell shape and symmetry. 1. Centric diatoms are non-motile, whereas many species of the pennate forms exhibit a gliding movement which is, in some way, dependent on the presence of a raphe. 2. Sexual reproduction of the Centrales is oogamous whereas that of the pennate species is generally isogamous. 3. Although species of the two groups sometimes inhabit the same regions, Centrales are more commonly planktonic and marine, whereas Pennales are usually freshwater forms and frequently occur as epiphytes and as components of the mud and soil flora.

The more detailed classification of diatoms depends almost entirely on the structure of the siliceous skeleton. The Centrales are divided into three major groups on the basis of cell shape and the presence or absence of particular processes. Genera such as *Coscinodiscus, Cyclotella* and *Melosira* are disc-shaped with no processes, whereas the valve surfaces of genera such as *Biddulphia* and *Chaetoceros* have various horns and bosses. A third group containing genera such as *Rhizosolenia* and *Corethron* also have valves with no processes, but the cells are elongated and have a complex girdle structure.

The classification of the pennate diatoms is based largely on extent of development of the raphe. *Tabellaria* and *Asterionella* are examples of the group of diatoms which only possess a pseudoraphe. Amongst the other forms it is possible to identify an increasing tendency for the development of a raphe; the valves of *Eunotia*, for example, show the beginnings of raphe development. *Achnanthes* and *Cocconeis* have a raphe on one valve only, but most genera (e.g. *Navicula*, *Bacillaria* and *Nitzschia*) have a raphe on each valve.

For the most part, examinations with the electron microscope have not changed the basic classification of the diatoms, but have modified the classification of individual species and genera. An example of this is *Phaeodactylum tricornutum*, formerly known as *Nitzschia closterium* forma *minutissima*. Its form is variable and it is known to exist in oval, fusiform and tri-radiate phases. Hendey[113] examined the fusiform and tri-radiate phases in the electron microscope and concluded that no evidence warranted its inclusion in the genus *Nitzschia*, and that there was also no evidence to suggest that it was a diatom. Later, however, Lewin[143] examined the oval phase in the electron microscope and observed a siliceous valve on one side only, and this valve had structures which suggested that the species was closely allied to *Cymbella*.

PHYLOGENY OF DIATOMS

Because of the siliceous skeleton an extensive fossil record of diatoms is known and a discussion of phylogeny based on the fossil evidence is possible with diatoms to a much larger extent than with other algae. Centric forms appear to be the most primitive, appearing first in the Jurassic, whereas the pennate forms do not make their first appearance until the Tertiary period. The evidence also suggests that the pennate forms were derived from the centric forms and that those pennate species with no raphe represent the early stages of this development (chloroplast structure and ecological habitat support this hypothesis).

9

PYRROPHYTA AND CRYPTOPHYTA

This chapter includes two algal divisions about which little is known. Several authorities consider the Dinophyceae and Cryptophyceae to be two classes in the same division, the Pyrrophyta; whereas others emphasise the difference between the two groups and create two divisions, the Pyrrophyta and the Cryptophyta. The two divisions are presented here in the same chapter for convenience, and not to suggest any close affinity between them.

Division PYRROPHYTA

Classes DESMOPHYCEAE and DINOPHYCEAE

This division is characterised by three main features: 1. Photosynthetic pigments include chlorophylls *a* and *c*, together with several carotenoids (e.g. dinoxanthin, and peridinin) unique to the Pyrrophyta. 2. The cells are biflagellate with one 'band-shaped' flagellum and another apparently of the acronematic type. The orientation and location of the flagella are also important diagnostic features and are discussed later in the chapter. 3. The detailed structure of the cell shows several characteristic features. For example, the nucleus is large and prominent and has a characteristic structure (see below); also most cells contain conspicuous membrane-bound *pusules* which bear a superficial resemblance to contractile vacuoles but show no contractile activity, and there are also distinctive sculpturing patterns on the cell surface.

Species of Pyrrophyta can be classified into two classes: the *Desmophyceae* and the *Dinophyceae* (earlier authorities usually recognised only one class, the Dinophyceae, but divided it into two subclasses, the *Desmokontae* and the *Dinokontae*), and they can be distinguished by three main features: the cell wall structure, the location of flagella and the form of the chloroplast.

Before discussing each class separately, the nucleus and mode of nuclear division is to be described, since it is essentially the same in both classes.

NUCLEUS AND MODE OF NUCLEAR DIVISION

A large distinctive nucleus has been observed in many species of Pyrrophyta, and, although many of the earlier observations gave conflicting reports on a number of details, Dodge[55] has surveyed the nuclear structure in 4 species of Desmophyceae and 9 species of Dinophyceae and has clarified much of the confusion. Most of the following generalisations are taken from the paper of Dodge.

In the interphase nucleus, the chromosomes remain in a condensed state, comparable with that found in the metaphase and anaphase condition of most dividing nuclei. Although a nuclear membrane is present in most species, Dodge could not observe such a membrane in *Prorocentrum micans* (Desmophyceae).

The early reports of the nucleus spoke of it as being 'moniliform' to indicate that the chromatin material appeared as a series of beads arranged along a thread. Such a moniliform appearance may be due to the condensed nature of the chromosomes at interphase combined with their helical structure (Dodge[56] has observed the helical structure of chromosomes of *Prorocentrum micans* and several of his photographs have 'beaded' (i.e. moniliform) appearance).

Division of the nucleus is mitotic, and longitudinal division of the chromosomes is now established. Observation of the pattern of anaphase movement and of the effect of various chemical and physical agents on division suggest that both centromeres and a spindle are absent. A nucleolus is usually present, and in *Oxyrrhis marina* and *Gyrodinium cohnii* it persists during division (cf. the endosome of the Euglenophyta, p. 125). However, in most species it behaves in the usual way, and no trace of it can be found between prophase and telophase.

Two pieces of evidence suggest a similarity between the structure of each individual chromosome of the Pyrrophyta and the nucleoplasm of blue-green algae. First, the chromosomes of

Prorocentrum micans contain deoxyribonucleic acid but not protein.[57, 58] In this they differ from the chromosomes of all other eucaryotic cells, where the DNA is combined with a special protein, histone; instead they resemble procaryotic cells (blue-green algae and bacteria) which do not have histones. Secondly, a change of fixative which produces a change in the electron microscopic image of the nucleoplasm of procaryotic cells (see p. 28) produces a similar change in the image of the individual chromosomes of the dinoflagellate *Amphidinium elegans*.[99] Thus, with a change in the calcium concentration the image of the chromosomes changes from a finely fibrillar nature to a more condensed appearance. (A fibrillar structure has also been observed in the chromosomes of *Noctiluca scintillans*).[2]

The nucleus of the Pyrrophyta has been discussed in detail because it is becoming increasingly apparent that details of cell structure can provide important criteria on which to base many discussions of algal evolution, and it is to be hoped that future investigations of these problems will be encouraged.

DESMOPHYCEAE

This class is small, its members are relatively rare, and all species are unicellular. The structure of the cell differs from that of the Dinophyceae in three ways: 1. The cell wall, when present, is not divided into a number of plates as in the Dinophyceae; instead it has a longitudinal suture dividing it into two *valves* (fig. 19 A). Transverse furrows are absent. 2. The two flagella are not situated laterally, but arise from the anterior of the cell. The thread-like flagellum is directed forwards and the 'band-shaped' one bends at right angles after emerging from the cell, and curves round the base of the other flagellum (fig. 19 B). 3. The Desmophyceae have chloroplasts which are usually plate-like (*Haplodinium*) or lobed (*Desmomastix*) and pyrenoids are usually present, whereas the Dinophyceae normally have numerous discoid chloroplasts without pyrenoids.

Reproduction of the Desmophyceae appears to occur by only one method, namely the longitudinal division of the cell when still motile. The division process shows many similarities with cell division of diatoms; for example, there is growth of the suture pushing the two cells apart, and after division each daughter cell retains one valve from the parent cell.

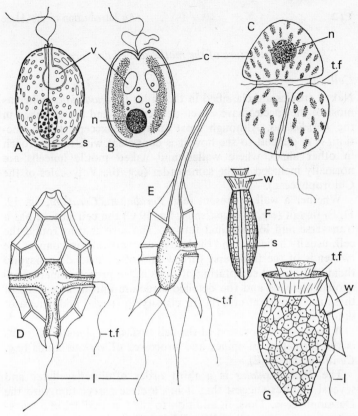

Fig. 19 Pyrrophyta

A, B, *Exuviaella marina*: A, side views howing suture dividing cell
into 2 valves; B, valve view. C, *Gymnodinium palustre*. D, *Peridinium
bipes*. E, *Ceratium cornutum*. F, G, *Dinophysis acuta*: F, side
view showing division into 2 valves; G, valve view. c, chloroplast;
l, longitudinal flagellum; n, nucleus; s, suture; t.f, transverse
flagellum; v, vacuole; w, wing. (A, B, F, G after Fritsch[90]; C, D, E
after Fott[84])

DINOPHYCEAE

This class is much larger than the Desmophyceae and the motile
unicellular forms (the dinoflagellates) are sometimes important
constituents of phytoplankton populations. Although motile
unicells form the bulk of the class a number of non-motile and
multi-cellular types also occur, and morphological developments
paralleling those in the Chlorophyceae and Xanthophyceae can
be identified.

Cell structure of the motile unicellular forms

Cell wall

Naked species are classified in the order *Gymnodiniales*, whereas most of the species have a well-defined wall and are classified in the *Peridiniales*. Although most authorities accept this distinction, it is difficult to see how it is consistent with the approach in other algae, where walled and naked motile unicells are normally included in the same order (e.g. the Volvocales of the Chlorophyceae).

Whether a wall is present (e.g. *Peridinium*, *Ceratium*, fig. 19, D, E) or not (e.g. *Gymnodinium*, fig. 19 C) the cell surface has a transverse and longitudinal furrow. The former runs round the cell, usually helically, and the latter runs vertically to connect the two ends of the transverse furrow and often extending beyond them. Additional sculpturing of the surface produces many hexagonal platelets and the detailed structure and arrangement of these are important criteria for classifying species and genera of dinoflagellates.

A common elaboration of the cell surface in planktonic forms is the formation of spines and processes of various kinds (e.g. *Ceratium*, fig. 19 E).

The *Dinophysidales* is a third order of dinoflagellates and several features suggest that it has a close association with the Desmophyceae. Thus, in addition to the longitudinal and transverse furrows the cell wall is divided vertically into two valves. The members of the Dinophysidales can also be recognised by another feature; the margins of the transverse furrow are frequently expanded as wing-like structures (fig. 19 F, G).

Flagella

The transverse 'band-shaped' flagellum emerges from the cell at the upper end of the transverse furrow and runs round the cell inside the furrow. The point in the vertical furrow at which the thread-like flagellum arises is variable and it trails behind the cell. The mechanism of flagella movement in the Dinoflagellate *Ceratium* has been investigated by Jahn *et al.*[124] A sine wave appears to pass from the base to the tip of the longitudinal flagellum, and such a movement presumably produces forward movement of the cell. The transverse flagellum has a helical wave passing along it and this produces a current in the water along the axis of the flagellum. Jahn and his co-workers also showed that when such a flagellum undulates in a groove, water currents are also produced perpendicular to the axis of the flagellum. That is, passage of a wave along the flagellum in the transverse groove might influence the forward movement of the cell.

Chloroplasts

When present, the chloroplasts are of variable form. For example, although many peripherally arranged discoid chloroplasts are common, some species of *Peridinium* (e.g. *P. umbonatum*) have a single lobed, axile chloroplast, and unlike the chloroplasts of most dinoflagellates they possess pyrenoids.

Colourless species are common in several genera (e.g. *Gymnodinium*, and the colonial *Polykrikos*). The nutrition of these colourless species is variable; although some are saprophytic, a holozoic mode of nutrition is common. This latter method appears to be well established in the colourless species *Gyrodinium hyalinum*, and also occurs in several holophytic species of *Gymnodinium* and *Ceratium*. A series of marine parasitic forms (ectoparasites such as *Oodinium* and *Apodinium*, and endoparasites such as *Blastodinium* and *Schizodinium*) are classified in the family *Blastodiniaceae*. Many of the *zooxanthellae* present in the tissues of marine invertebrates produce *Gymnodinium*-like swarmers[246] and Freudenthal[87] suggests that all zooxanthellae are members of the Blastodiniaceae.

Eye-spot (*stigma*)

An eye-spot is generally found near the base of the longitudinal flagellum of most naked freshwater species, but is absent from marine species and from all species with a cell wall (the simple walled genus *Glenodinium* is an exception).

Cells of one group of marine forms (e.g. *Nematodinium* and *Erythropsis*) contain an *ocellus*. This is assumed to be an elaborate form of eye-spot, although there is no conclusive evidence for its function. It consists of a colourless refractive *lens* partially buried in a pigmented mass, the *melanosome* (fig. 20 C).

Range of vegetative structure

The above discussion of cell structure has been concerned solely with the dinoflagellates and the three orders *Gymnodiniales* (naked forms), *Peridiniales* (armoured forms) and *Dinophysidales* (armoured forms with two valves) have been recognised. However, the motile unicell is not the only vegetative type in the Dino-phyceae and other orders can be identified, the *Rhizodiniales*, *Dinococcales*, *Dinocapsales* and *Dinotrichales* representing the amoebid, coccoid, palmelloid and filamentous forms respec-tively. Thus the class shows morphological types analogous to many found in the Chlorophyta, Xanthophyta and Chrysophyta.

Motile colonial forms are rare and are classified as members of the three dinoflagellate orders mentioned above. For example, *Polykrikos* consists of 2, 4 or 8 *Gymnodinium*-like cells united to form a chain (fig. 20 B). Also, individuals of *Ceratium* sometimes unite to form temporary colonies with the spine of one cell fixed in the groove of an adjacent cell.

Species which are non-motile for the most of their life-history are usually interpreted as being *coccoid*, and all possess a cell wall but the surface is not sculptured into a number of platelets. The usual method of reproduction is by the formation and liberation of a number of zoospores. The daughter cells of *Hypnodinium* sometimes fail to acquire flagella and are liberated as aplanospores.

Palmelloid forms (*Dinocapsales*) are found in only two genera, *Gloeodinium* and *Urococcus*. *Gloeodinium* is the more common and consists of 2, 4 or 8 relatively large cells embedded in a gela-tinous mass which has recognisable striations near the surface. The aggregations are formed by failure of daughter cells to separate after division, and they rarely exceed the 8-celled stage. Asexual reproduction occurs by the formation of *Gymnodinium*-like zoospores from each of the vegetative cells.

Forms which are permanently amoeboid are classified as a separate order, the *Rhizodiniales*. As in the amoeboid forms of the Xanthophyta and Chrysophyta, the nutrition is frequently holozoic. Reproduction of such forms is usually achieved by the production of *Gymnodinium*-like zoospores within a temporary cyst.

There are two filamentous genera, *Dinothrix* (fig. 20 A) and *Dinoclonium*. The filament increases in length by vegetative cell division at right angles to the axis of the filament, and the daughter cell initially has a transverse furrow as in motile unicells, but this is later lost as the cell matures into a vegetative cell of the fila-ment. The organisms reproduce by the production of *Gymno-dinium*-like zoospores (either singly or in twos) from each vegetative cell (fig. 20 A).

Most non-motile forms described above reproduce asexually

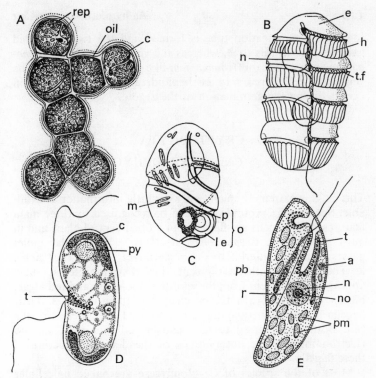

Fig. 20 Pyrrophyta and Cryptophyta

A, *Dinothrix paradoxa*. B, *Polykrikos kofoidi*. C, *Nematodinum armatum*. D, *Hemiselmis rufescens*. E, *Chilomonas paramaecium*, diagram of electron micrograph. a, amphosome; c, chloroplast; e, epivalve; h, hypovalve; le, lens; m, nematocyst; n, nucleus; no, nucleolus; o, ocellus; oil, oil droplets; p, pigmented body of ocellus; pb, parabasal body; pm, paramylum; py, pyrenoid; r, rhizoplast; rep, cells dividing to form *Gymnodinium*-like swarmers; t, trichocyst; t.f, transverse flagellum. (A after Fott[84]; B, C after Fritsch[90]; D after Parke[188]; E after Anderson[5])

by the production of zoospores, and all of these show a close resemblance to the vegetative cells of *Gymnodinium*. This fact has generally been interpreted as evidence that a *Gymnodinium*-like organism represents the starting point for the evolution of the Dinophyceae.

Reports of sexual reproduction have been described for various species, but the process appears to be established in only two genera, *Ceratium* and *Glenodinium*. In *Ceratium* two cells come together, a conjugation tube is formed and the two amoeboid gametes fuse in this tube, whereas the vegetative cells of *Glenodinium* appear to form *Gymnodinium*-like gametes which fuse. The vegetative cell appears to be haploid, and reduction division occurs during the germination of the zygote.

Division CRYPTOPHYTA

Class CRYPTOPHYCEAE

This class is extremely small, and very little information is available. Species of Cryptophyceae can be recognised by three main features. 1. The cells are flattened in a dorsi-ventral plane and in most genera (e.g. *Cryptomonas*, *Rhodomonas*) a tubular gullet passes from the anterior to the posterior of the cell (fig. 20 E). 2. Their photosynthetic pigments include biloproteins; these appear to be different from those of the blue-green and red algae, but as yet they have not been characterised in detail. 3. Their flagella are usually equal and are located at the anterior of the cell. Both are referred to as 'band-shaped' but there are no electron microscopic observations of the detailed structure of these flagella.

Most of the species of Cryptophyceae are naked unicellular motile forms. Although details of cell structure are consequently important to an understanding of the group, members have been largely neglected in recent electron microscopic investigations of algae, so that most of the following remarks come from older work, and confirmation or modification is needed.

CELL SHAPE

As mentioned above, the cells are normally flattened in a dorsi-ventral plane, but the group shows an interesting modification of this form. Genera such as *Protochrysis* and *Hemiselmis* have kidney-shaped cells with the flagella located laterally (fig. 20 D); that is, the cells bear a superficial resemblance to motile stages of the Phaeophyceae.

Flagella

The two equal (sometimes slightly unequal) flagella are usually located anteriorly, except in the genera mentioned in the previous paragraph. The 'band-shaped' nature of the two flagella has not been investigated with the electron microscope, but the presence of rhizoplasts connecting one of the basal granules to the nucleus in *Chilomonas* has been confirmed by a recent electron microscopic investigation of this genus. The rhizoplasts are fibrous strands passing from one of the basal granules to the posterior of the cell and the fact that it touches the nuclear membrane at one point appears to be incidental (fig. 20 E). Anderson[5] interprets this rhizoplast as the remains of a flagellar rootlet similar to that found in the ciliates.

Nucleus

The cells are uninucleate and the nucleus is located towards the posterior of the cell. It appears to have a well-defined membrane and a prominent nucleolus.

Chloroplasts

Two large parietal chloroplasts are present in most cells, although the cells of *Cryptochrysis* and *Cyanomonas* contain numerous small discoid chloroplasts. Pyrenoids appear to be absent, although pyrenoid-like bodies are distributed throughout the cell. Electron microscopic examination of *Chilomonas paramaecium* confirms the presence of spherical electron opaque homogeneous structures which are presumably equivalent to the supposed pyrenoids of earlier light reports (fig. 20 E). Anderson[5] calls these structures *amphosomes* and suggests that they might be vestigial eye-spots or pyrenoids.

Other cytoplasmic structures include a single contractile vacuole normally present in the anterior part of the cell, Golgi apparatus and mitochondria.

Trichocysts

These have also been observed in the cells of *Chilomonas* (fig. 20 E). They appear in the electron microscope as opaque smooth angular bodies with a complex internal structure. They are frequently extruded from the cell and the alternative name of *ejectisome* has been suggested.[5] Such trichocysts have also been found in the dinoflagellate *Oxyrrhis marina* where the trichocyst structure resembles that found in *Paramaecium*. There is no evidence on their chemistry or their function (if any), but Anderson suggests that they originate at the Golgi apparatus.

REPRODUCTION

The usual method of reproduction of the motile unicellular forms is by longitudinal division of the cell when still motile. In addition, some species produce thick-walled cysts and in *Cyanomonas* and *Cryptomonas* the resting cell becomes embedded in mucilage and the daughter cells remain together as temporary palmelloid stages.

In two genera, *Phaeococcus* and *Phaeoplax*, the palmelloid condition is the normal vegetative phase, and reproduction is effected by the production of biflagellate swarmers. There is one other algal development in the Cryptophyceae. *Tetragonidium* is a *coccoid* species consisting of tetrahedrally arranged cells each of which has a well-defined wall. This genus has been identified as a member of the Cryptophyceae by its *Cryptochrysis*-like swarmers.

Although pigmentation might suggest that the Cryptophyta represents an intermediate stage in the evolution of eucaryotic forms from procaryotic algae,[4, 59] the above brief discussion illustrates that there are no structural features to support this idea.

I O

EUGLENOPHYTA: EUGLENOPHYCEAE

Division EUGLENOPHYTA
Class EUGLENOPHYCEAE

This group of organisms shows a very limited range of vegetative forms, since, with the exception of the genus *Colacium*, which forms palmelloid or dendroid colonies, all euglenoids are unicellular flagellated organisms. Species of the Euglenophyceae are characterised by four main features. 1. Their photosynthetic pigments resemble those of the Chlorophyceae since they contain chlorophylls *a* and *b*; carotenoids such as β-carotene, lutein, neoxanthin, astaxanthin[103] and antheraxanthin[102] are also present. 2. All euglenoid cells are naked. 3. They have one, two or three flagella (all of the pantonematic type) arising from an invagination at the anterior of the cell. 4. The reserve product is a polysaccharide, paramylum.

DETAILS OF CELL STRUCTURE

Cell envelope

Euglenoid cells do not possess a cell wall; instead they are bounded by a cell membrane termed a *pellicle* or *periplast*. In genera such as *Phacus*, *Rhabdomonas* and *Menoidium* the pellicle is rigid and the cell has a fixed shape, whereas in other genera (e.g. *Euglena* and *Distigma*) it is flexible and in some species the cell changes shape continually (e.g. *Euglena gracilis*). In the two species which have been examined most extensively in the electron microscope (*Euglena gracilis*[95] and *Peranema tricophorum*) the outer surface is highly convoluted. In *Peranema tricophorum* the pellicle consists of two unit membranes, an outer one consisting of two electron opaque layers separated by a less opaque area and the inner cytoplasmic membrane showing the same basic structure. Thus, in this species the pellicle is *not* equivalent to the cytoplasmic membrane. In *Euglena gracilis*, the presence of two unit membranes is less well established, although Gibbs[95] has observed that the outer electron opaque layer of a unit membrane can sometimes be distinguished into two opaque layers separated by a less opaque area.

In both *E. gracilis* and *Peranema tricophorum* a system of filaments, or tubules, has been identified beneath the pellicle. In *Peranema* an additional system of heavy fibres alternates with the filaments. Roth speculates that these fibres may be concerned in the rhythmic changes in shape accompanying the gliding movement of *Peranema*. Leedale argues that because these fibres cannot be found below all crests in the convoluted pellicle, they are unlikely to have a contractile function.

More recently, Leedale[141] has examined the pellicle structure of three other species of *Euglena*, *E. spirogyra*, *E. spirogyra* var *fusca* and *E. pisciformis*, and this paper is recommended as a critical review of pellicle structure in euglenoids. Some interesting points which emerge from this paper are as follows:

(a) Examination with the light microscope under suitable conditions shows the surface to have helical striations and that these are almost always left-handed (anticlockwise from the posterior of the cell)

(b) Electron microscopic examination confirms that the pellicle has strips of material passing along the cell in a helical manner. These have an elaborate shape in cross-section

(c) Some strips are ornamented with a series of knobs

(d) During 'euglenoid movement' the cell shape changes and this implies that the pellicle strips are flexible and elastic. Examination confirms that there is a certain amount of relative movement between strips

(e) A possible relevance to the elastic nature of the pellicle is the fact that the pellicle is composed largely of protein and this may be of the fibrous elastic type.

In four genera (*Trachelomonas*, *Strombomonas*, *Ascoglena* and *Klebsiella*) a special envelope (lorica) surrounds the cell and is separated from the protoplast by a well-marked space. The lorica can be rigid or gelatinous, but no trace of cellulose has yet been detected.

Gullet and contractile vacuolar system

A characteristic feature of the euglenoid cell is the presence of a flask-shaped invagination at the anterior of the cell (fig. 21). This has been termed the *gullet* and consists of an enlarged basal *reservoir* and a narrow *cytopharynx*, and the reservoir connects with the interior of the cell at the *cytostome*. The name 'gullet' arose from obvious associations between the structure and ingestion of solid particles. Although it is clear that colourless forms such as *Peranema*[24] ingest solid particles there is no evidence for such ingestion by pigmented cells. Members of the Peranemaceae (see p. 127) contain two pharyngeal rods which occur near the gullet and are orientated parallel to it (fig. 21 C). These rods have been implicated in the ingestion of solid particles but no definitive evidence on their possible mechanism of action is available.[122, 123]

A contractile vacuole usually occurs near the reservoir. The vacuole empties into the reservoir and a new contractile vacuole

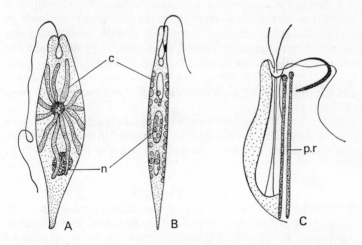

Fig. 21 Euglenophyta

A, *Euglena viridis*. B, *E. acus*. C, *Heteronema* (anterior end showing gullet and pharyngeal rod apparatus). c, chloroplast; n, nucleus; p.r, pharyngeal rods. (A, B after Fritsch[90]; C after Jahn[122])

develops from the fusion of many smaller vacuoles. The suggestion that contractile vacuoles have some osmotic regulatory function is supported by the fact that the rate of pulsation decreases when the osmotic pressure of the external medium is increased.

Flagella

Euglenoid cells possess one (e.g. *Euglena, Lepocinclis*), two (*Heteronema, Entosiphon*) or three (*Euglenamorpha*) emergent flagella arising from the posterior of the reservoir and passing through the cytopharynx. All flagella are of the pantonematic type, and in those species possessing two flagella, one normally trails behind the cell and the other is generally directed forwards (*Eutreptia* is an exception since the two flagella appear to be directed forwards). The basic mode of movement of the euglenoid flagellum appears to resemble that of the flagellum of other algal groups in showing the passage of a wave from the base to the tip of the flagellum. If the flagellum is anteriorly directed in the euglenoids it is difficult to envisage how the mechanism of movement can correspond with that found in other algal groups.

A swelling of *paraflagellar body* near the base of the flagellum is believed to be the photo-receptor in the photostatic response of euglenoids. However, despite the fact that in several metazoa photoreceptors are often derived from cilia,[173, 199] no conclusive evidence for the photoreceptive function of the paraflagellar body is available; certainly the critical steps of isolating the structure and identification of a photosensitive pigment has not yet been performed.

At the base of the flagellum is the basal granule, or blepharoplast. This structure has long been implicated in nuclear division, and this aspect will be discussed later in the chapter (see p. 125).

Stigma or eye-spot

In the light microscope, the eye-spot or stigma appears as a red-orange spot situated near the base of the flagellum. In the electron microscope the stigma of *E. gracilis* appears as a curved plate of many orange-red granules. According to Gibbs[95] these granules appear to be loosely arranged, thus disagreeing with Wolken[244] who showed close hexagonal packing of the granules. Wolken[244] assumes the stigma to consist of carotenoid molecules arranged in a monolayer. The most widely accepted function of this stigma is that of a light-absorbing shield, which, depending on the orientation of the organism, prevents light from reaching the paraflagellar body, the true photoreceptor. Evidence for this comes from observations of colourless organisms. Although the stigma is present in all pigmented cells, it is absent from some colourless forms, and early observations indicated that only those colourless forms containing a stigma showed phototactic responses. Further evidence that a close association between stigma and the flagellar swelling may function in some way in photorection comes from analogy with several other algae. Thus, in *Fucus serratus* spermatozoid[159] the posterior flagellum is swollen and adpressed to the membrane immediately above the stigma (see fig. 24 A, p. 139). Similarly in *Chromulina psammobia*[207] there is a short swollen internal flagellum lying in the cup of the stigma. Although the role of the stigma as a shield in phototaxis of *Euglena* is widely accepted, this organelle does not have such a function in all groups since species of *Chlamydomonas* (see p. 41) which lack a stigma exhibit prototactic responses.

Chloroplast

The photosynthetic pigments of coloured euglenoids are contained in well-defined membrane-bound chloroplasts of variable form, and the form of chloroplast is an important criterion in the classification of species of *Euglena*.

Because of the remarkable ease with which cells can be made to lose their chloroplasts by changing environmental conditions, euglenoids have been an attractive experimental material for the investigation of those factors involved in the formation of chloroplasts. Most of the work has been undertaken with the species *Euglena gracilis*, and although therefore it is difficult to assess how generalised the picture might be, the observations will be presented in some detail since the investigations have formed such a major part of studies concerned with the Euglenophyceae.

In cells grown in the light under autotrophic conditions, electron microscopic examination reveals the chloroplasts to contain a granular matrix traversed by 10–45 moderately dense bands. Earlier investigators[238, 245] referred to these bands as lamellae but Gibbs[95] has shown each band to consist of two, three, four or five closely adpressed discs. Gibbs emphasised the similarities between this and the chloroplast structure prevalent in other classes of algae.

When *Euglena* is grown in darkness with organic carbon

sources, synthesis of photosynthetic pigments is prevented, and the cells do not contain chloroplasts. Instead the cells contain a large number of vacuoles and paramylum granules. One point of controversy has been over the question of whether such cells contain *proplastid* elements. The fact that chloroplasts are formed within a few hours of transferring such dark-grown cells to the light suggests that proplastids are present, and this has been confirmed by microscopic examination of cells under conditions when fluorescence by protochlorophyll can be detected.[74, 97] Further details of chloroplast formation can be found in a series of papers by Epstein, Schiff and their colleagues.[9, 73, 151, 215, 216, 229, 230] The transference from darkness to illuminated conditions is also accompanied by interesting chemical changes. For example, in addition to synthesis of the photosynthetic pigments, Brawerman[16] has identified the formation of specific ribosomes believed to be necessary for the synthesis of chloroplast material.

Bleaching of *Euglena* (accompanied by the loss of chloroplasts) can also be caused by a variety of physical and chemical agents, including treatment with streptomycin, ultraviolet light, elevated temperature and several organic carbon sources. One particularly important result has been the finding that inactivation of green colony formation by *Euglena* caused by ultraviolet light is irreversible (i.e. heritable effect), non-lethal (i.e. cytoplasmic) and has an action spectrum with a peak of 260–80 μ. These results suggest that *Euglena* has self-reproducing cytoplasmic entities which contain nucleoprotein and which are responsible for chloroplast formation.[132, 151]

In some species proteinaceous pyrenoids are present in the chloroplasts; these appear in the light microscope as spherical bodies of different staining properties from the remainder of the chloroplast. The fine structure of the pyrenoid as revealed by the electron microscope shows a granular matrix which is denser than the matrix of the remainder of the chloroplast and traversed by bands corresponding to those of the chloroplast. The deposition of paramylum usually occurs at the pyrenoid, although in many species paramylum bodies are found in the cytoplasm completely separate from the chloroplast.

NUCLEUS AND NUCLEAR DIVISION

Many of the earlier observations on the nucleus and mode of nuclear division gave conflicting evidence on a number of features, but in 1958 Leedale[139] examined 24 euglenoid species from 13 genera and clarified much of the confusion. The description below is therefore a summary of Leedale's paper.

The euglenoid nucleus contains one or more deeply staining granules, *endosomes*. In general appearance they appear homologous to the nucleoli of many other nuclei. However, their behaviour differs from that of the nucleolus in that they persist during mitosis and retain their ribonucleic acid (RNA). Other authors have suggested an homology between endosomes and 'centrioles' or 'centrosomes'. However, this does not appear to be valid and from the results of Leedale's survey, centrioles do not appear to be part of the euglenoid nuclear machinery; rather, the endosome appears to be a passive body, dividing at the time of nuclear division but not controlling the division process.

Various reports have suggested a close relationship between the flagellar apparatus and the nucleus; in particular that the blepharoplasts function as division centres and that in some cases a system of fine strands (*rhizoplasts*) connects the basal granules to the endosome or to a 'centrosome' in the nuclear membrane. However, despite the fact that the nucleus moves towards the anterior of the cell immediately before division, Leedale could find no evidence relating the blepharoplasts to the nucleus.

A spindle appears to be absent (e.g. colchicine has no effect and the movement of the chromosomes at anaphase appears to be staggered) and the division appears to be truly mitotic with longitudinal duplication of each chromosome. This latter observation does not support earlier suggestions that the chromatin material of euglenoids was present as a continuous spireme which broke transversely at the time of division. Meiosis is generally thought to be absent from the euglenoids but Leedale[140] has published photographs apparently showing meiotic figures.

NUTRITION

The relative ease with which euglenoids can be made to lose their photosynthetic pigments and associated structures has resulted in heterotrophic aspects of their nutrition being emphasised. Thus, *Euglena* species are facultative heterotrophs, growing phototrophically with carbon dioxide as sole carbon source when illuminated but growing heterotrophically when supplied with a variety of organic carbon sources. Other euglenoids exhibit holozoic nutrition. Although some of the colourless species (e.g. *Peranema*) ingest solid particles, a holozoic mode of nutrition is probably less widespread amongst the euglenoids than was once thought, and coloured species probably never show it.

REPRODUCTION

The normal method of reproduction is by longitudinal division of the motile cell. In uniflagellate species the original parent flagellum is either lost or transferred to one daughter cell, the other acquiring a new flagellum.

The formation of cysts occurs in forms such as *Euglena*, *Phacus* and *Trachelomonas*. Also, species of *Euglena* occasionally form palmelloid stages consisting of non-flagellated organisms embedded in a gelatinous matrix. This stage is, however, temporary and no vegetative division occurs during the palmelloid phase. This contrasts with the condition found in *Colacium*. In this genus a palmelloid or dendroid colony is the normal vegetative form, flagellate stages are rare, and vegetative division is confined to the palmelloid stage.

The few reports of sexual reproduction are unconfirmed and there is no conclusive evidence for its occurrence in the Euglenophyceae.

CLASSIFICATION AND RELATIONSHIP OF THE EUGLENOPHYCEAE

Euglenoids are divided into four families: Euglenaceae, Astasiaceae, Colaciaceae and Peranemaceae. The *Euglenaceae* include all pigmented forms together with the colourless forms immediately derived from them. It concludes such genera as *Euglena*, *Phacus*, *Trachelomonas*, *Klebsiella*, *Euglenamorpha* and *Strombomonas*.

Colaciaceae was suggested by Smith[221] to accommodate the single genus *Colacium*, and most authorities recognise that the distinctive features of the genus (outlined above) warrant the creation of a separate family.

The colourless forms (except those immediately derived from pigmented cells) are included in two families, *Astasiaceae* and *Peranemaceae*, and a number of criteria are used to distinguish between them. Astasiaceae are generally considered free swimming, saprophytic organisms with a single flagellum and do not possess pharyngeal rods. Whereas members of the Peranemaceae generally show a gliding form of motility, they are holozoic, possess two flagella and pharyngeal rods. Jahn[122, 123] has emphasised that the above distinctions are not absolute and, depending on which criteria are used, various genera can be classified into one or other of the families.

It was pointed out in the introductory chapter (see p. 10) that the strongest reason for including flagellates in a study of the algae was their obvious close affinities with 'algal' types. In the Euglenophyta, only two features indicate algal affinities; first, the presence of the palmelloid genus *Colacium*, and secondly, the possession of photosynthetic pigments. However, certain aspects of the pigmentation suggest affinities with animal groups. Thus, keto-carotenoids have been identified in *Euglena gracilis* and *Trachelomonas volvocina* and the role of such compounds in the biosynthesis of astaxanthin from β-carotene has been established in several animal systems, whereas the biosynthesis of carotenoids in plant groups does not involve such keto-carotenoids as intermediates.[103]

I I

ALGAE OF

UNCERTAIN SYSTEMATIC POSITION

This chapter includes two classes, the Chloromonadophyceae and the Prasinophyceae, both of which have very few species. More specimens must be examined before they can be included in one of the existing phyla or before they are separated as well-defined, circumscribed phyla themselves. At the end of the chapter the single genus *Cyanidium* is also described since it appears to occupy a unique position in the algae.

CHLOROMONADOPHYCEAE

This class includes very few genera (e.g. *Monomastix*, *Merotrichia*, *Vacuolaria*, *Raphidomonas*) and all are motile unicells which have no cell wall. Instead of a wall, the cells have a soft periplast and they consequently show considerable changes of shape. The cells are usually flattened dorsiventrally, they normally have a longitudinal groove on the ventral surface and some genera (e.g. *Trentonia*, *Gonyostomum*) have a cavity at the anterior of the cell comparable to a gullet (fig. 22 A, B). Apart from the uniflagellate *Monomastix* all genera are biflagellate and the flagella usually arise from a depression at the anterior of the cell, one trailing behind the cell and one directed forwards. There are few published details of cell structure and no electron microscope investigations appear to have been made. There appears to be no eye-spot, but contractile vacuoles are normally present; some species have trichocysts. The cells normally have many discoid chloroplasts but there appears to be no information on the nature of the pigments.

The relationship of the chloromonads to other algae is difficult to establish. Features such as the naked, dorsiventral cells with longitudinal furrows, a gullet and trichocysts suggest an affinity with the Cryptophyta. However, the flagella of the two groups are different and, unlike the Cryptophyta, cells of the Chloromonadophyceae appear to accumulate oil as the storage product. However, more information on the pigments and possibly on the fine structure of the cells is needed before their relationship to other algae can be assessed.

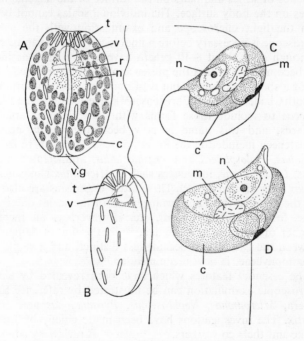

Fig. 22 Chloromonadophyceae and Prasinophyceae

A, B, *Gonyostomum ovatum*: A, ventral view; B, lateral view. C, D, *Micromonas squamata*: C, stationary cell with flagellum curved round the cell in characteristic position; D, cell swimming with flagellum in characteristic position; flagellum originates at anterior of cell. c, chloroplast; l, lipid bodies; m, mitochondria; n, nucleus; r, reservoir; t, trichocysts; v, vacuole; v.g, ventral groove. (A, B after Fott[84]; C, D after Manton and Parke[165])

E

PRASINOPHYCEAE

Most species of this class used to be included in the Chlorophyceae, Xanthophyceae or Chrysophyceae, but increasingly
intensive investigations with the electron microscope argue for
their separation from these classes.

The essential feature distinguishing members of this class is the
presence of scales and hairs on the surface of the flagella, and of
scales on the body surface. The individual scales cannot be seen
with the light microscope and examination with the electron
microscope is necessary before a final identification can be made.
Although only about 8–10 genera have so far been examined, it
has become evident that the presence of flagellar and body scales
is not a specific difference but is of major phyletic importance.

Both Chadefaud[19] and Christensen[26] suggested the name *Prasinophyceae* to include those forms with scaly flagellar and body
surfaces, and this name is now becoming widely accepted.
Christensen included in the Prasinophyceae such genera as *Pyramimonas*, *Halosphaera* and *Prasinocladus*, and created another
class, *Loxophyceae*, for genera such as *Micromonas* and *Bipedino-
monas* (now *Heteromastix*). These latter organisms are also scaly
but the detailed form and arrangement of the scales differs from
those found in *Pyramimonas*, etc. However, as an increasing
number of genera was examined,[168, 192] the line of demarcation
between the two classes became less distinct, and a single class,
Prasinophyceae, is now recommended.[191]

The essential features which have been revealed by electron
microscopic examination can be illustrated by reference to four
genera, *Micromonas*, *Nephroselmis*, *Pyramimonas* and *Heteromastix*. The investigations have been made chiefly by Manton,
Parke and their co-workers.[165, 167, 168, 192] A review by Manton[158]
is also recommended.

The cells of *Micromonas* (fig. 22 C, D) are small and are difficult to observe in the usual plankton sample. The single flagellum
arises laterally (it is posterior in *Pedinomonas*) and is covered by
hairs and a single layer of scales. In some species the surface of
the cell is also covered by a single layer of scales and the form
of each scale is similar to that of the flagellar scales.

The cells of *Pyramimonas* are larger than those of *Micromonas*.
They are heart-shaped with four flagella arising between lobes at

the anterior of the cell (fig. 23 A). Unlike *Micromonas* there is more than one layer of scales on the flagellar and body surfaces. For example, the body is covered by an inner layer of very small scales, a middle layer of larger scales and an outer layer of large net-like scales. Moreover, the flagellar scales are different from those on the body of the cell. The flagella also have a covering of fine hairs.

It was this much greater complexity of *Pyramimonas* (compared

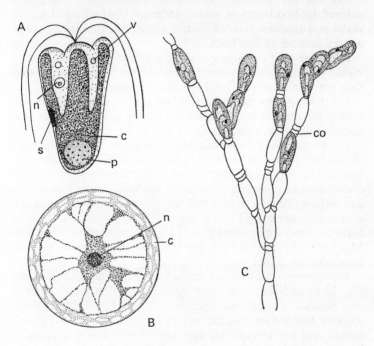

Fig. 23 Prasinophyceae

A, *Pyramimonas grossii*, light microscope view. B, *Halosphaera viridis*, optical section showing numerous discoid chloroplasts (c) in living cytoplasm and nucleus (n) suspended by cytoplasmic strands. C, *Prasinocladus subsalsa* part of colony. c, chloroplast; co, collar formed by overlapping cell walls; n, nucleus; p, pyrenoid; s, stimga; v, vacuole. (A after Parke[188]; B, C after Fritsch[90])

with *Micromonas*) which supported the idea of two separate classes. However, Parke and Rayns[192] examined *Nephroselmis gilva* with the electron microscope and found several features suggesting that the organism bridges the gap between *Micromonas* and *Pyramimonas*. Thus, although *Nephroselmis* has only one layer of scales, the form of the flagellar scales (a rounded or angular scale with a ridge of adnate spine and their arrangement (nine longitudinal rows) are similar to *Pyramimonas*.

The flagellar and body surfaces of *Heteromastix*[168] are also covered by two layers of scales: an inner layer of small square scales and an outer layer of stellate forms. The flagella also have a large number of fine hairs.

For some time the origin of these elaborate scales remained a mystery, but sufficient evidence has now been gained to show that they are formed at the Golgi apparatus (dictyosomes).

The four genera mentioned above are all unicellular motile forms; the class is not composed solely of such forms but also contains the genera *Prasinocladus* and *Halosphaera*.

Prasinocladus is a dendroid colony formed in the following way: the motile stage attaches itself to the substratum by the anterior end, and the flagella are lost to produce the sedentary phase. Later the protoplast contracts away from the wall, forms a new wall and acquires flagella. It may now escape as a zoospore, although it is sometimes caught at the aperture and germinates there to leave an empty stalk behind (fig. 23 C). Recent work by Parke and Manton[191a] has revealed the fine structure of the motile stage as being similar to that of *Pyramimonas*.

Halosphaera consists of large spherical non-motile cells (fig. 23 B) and they reproduce by the formation and liberation of a number of zoospores. Examination with the electron microscope[168] has shown that the motile stage is similar to *Pyramimonas*, and the demarcation between the two genera appears to be in doubt.

Thus details of flagellar structure and of the body surface support the creation of a separate class. Other features diagnostic of the class might include a peculiar form of starch (starch shells not stainable with iodine are frequently found), and a special type of fibrous band joining the flagellar bases to the plastid or nuclear surfaces.[168] Some of the more exciting work of the future will probably include biochemical analyses (e.g. photosynthetic pigments, food storage products, etc.) of these organisms in an attempt to relate them to other algae.

CYANIDIUM

Cyanidium caldarium is a widespread alga with a gross morphology and a developmental history similar to that of *Chlorella*. However, there are several anomalous features, and it cannot be assigned to any algal division with certainty. In the past it has been described as a blue-green alga, a coccoid cryptomonad, a green alga, a red alga and an anomalous chlorophyte.

Cyanidium is important because it appears to have some features in common with blue-green algae, and others with eucaryotic algae. It is therefore relevant to a discussion of the possible evolution of eucaryotic forms from procaryotic algae.

It resembles blue-green algae in having *c*-phycocyanin and like blue-green and red algae, its utilisation of light absorbed by the biloproteins is more efficient than its utilisation of light absorbed directly by chlorophyll *a*.[3] However, it differs from blue-green algae in several important ways. In particular, it lacks myxoxanthophyll[101] and examination with the electron microscope confirms that the cell is more highly differentiated than that of blue-green algae. Thus, with the electron microscope one can observe a well-defined nucleus, a membrane-bound chloroplast, vacuoles, mitochondria and an endoplasmic reticulum.[3, 172] The presence of a biloprotein, together with a eucaryotic type of cell architecture, might suggest that it is a coccoid cryptomonad (cf. ref. 81). However, the phycocyanin is different, chlorophyll *c* is not present and the main carotene is *β*- and not *α*-carotene. Because of these facts, Allen[3] considers the alga to be an anomalously pigmented chlorophyte, but Mercer et al.[172] emphasise the differences between the cell structure of *Cyanidium* and that of *Chlorella*.

Two features of cell structure are of relevance to its possible phylogenetic status. First, the membrane surrounding the chloroplast is sometimes indistinguishable from the lamellae, and secondly, the bands traversing the chloroplast consist of only one disc (i.e. two lamellae). The second feature contrasts with most algae, where the bands normally consist of more than one disc, although the red algae (Rhodophyta) also have bands of a single disc (see p. 149).

PHAEOPHYTA: PHAEOPHYCEAE

Division PHAEOPHYTA

Class PHAEOPHYCEAE (Brown algae)

Apart from a few freshwater genera (e.g. *Pleurocladia, Bodanella, Heribaudiella*) all brown algae are restricted to the sea, and are prominent in the littoral zone of the shore. They are characterised by four main features: 1. The photosynthetic pigments include chlorophylls *a* and *c*, together with xanthophylls such as fucoxanthin and diatoxanthin. 2. Food storage products include laminarin and mannitol. 3. Characteristic constituents of the cell wall include alginic acid and fucinic acid. 4. The motile stages have a distinctive appearance, almost all being pear-shaped with 2 flagella (1 acronematic and 1 pantonematic) arising from the side of the cell.

The degree of morphological complexity found in the Phaeophyceae is greater than any reached by algae discussed in previous chapters. For example, whereas the heterotrichous form is the most elaborate type of plant body in the Chlorophyceae it is the simplest type in the Phaeophyceae. The methods of reproduction and the life-histories show a comparable high degree of specialisation.

CELL STRUCTURE

All cells of Phaeophyceae have walls with an inner cellulose layer and an outer gelatinous layer of pectic material containing characteristic mucilages such as alginic acid and fucinic acid. The cellulose was once considered to be identical to that in the walls of higher plants but Cronshaw et al.[37] have shown it not to consist solely of glucose but to resemble the cellulose of *Ulva, Enteromorpha* and all red algae (except *Porphyra*) in containing glucose and xylose as the sub-units. The wall structure of all brown algae examined by Dawes et al.[40] was also identical to that of the Ulvales (Chlorophyceae).

Most cells contain many small vacuoles, and small colourless highly refractive vesicles (*fucosan-vesicles*) can be observed in many brown algae. It is not known whether fucosan is a food reserve or represents a 'waste' by-product of metabolism, but their abundance appears to be correlated with sites of high metabolic activity, such as dividing cells.

Apart from the stellate axile chloroplasts of *Pilayella fulvescens*, all others appear to be parietal, and most cells of brown algae contain numerous discoid chloroplasts. The greatest variety of chloroplast form in vegetative cells occurs in the *Ectocarpales*, an order which shows the least elaboration and differentiation. In this order one can identify single (or a few) plate-like chloroplasts, as well as ribbon-shaped and discoid forms. The only detailed electron microscopic examination of the chloroplasts appears to be that of Manton[156] in the zoospore of *Scytosiphon lomentarius*. It is bounded by a well-defined membrane and is traversed by a number of bands ('lamellae'). There are no grana-like structures but the number of discs per band cannot be estimated from the micrographs.

The cells contain a prominent membrane-bound *nucleus* which divides mitotically and which contains one or a few nucleoli. Although the resting nuclei do not normally appear to contain chromatin, those of several algae (particularly *Halidrys siliquosa*[179, 205] and species of *Cystoseira*[205, 210]) contain many Feulgen-positive bodies of about 0.6μ diameter. These are the *chromocentres* which, as the chromosomes differentiate at prophase, can be seen to be arranged along the chromosomes, but later, during condensation of the chromosomes, they become indistinguishable. The number of chromocentres is variable and bears no relation to the cytological state (i.e. diploid or haploid) of the cell.[205] These structures resemble those of some Angiosperms, but, although they appear to be found in no algal group except the Phaeophyceae, their significance (if any) is unknown. The nuclei of Phaeophyceae afford the clearest example amongst algae of *centrosomes* at the polar foci of the dividing nucleus.

The fine structure of some motile cells

Cells of the mature vegetative thallus of Phaeophyceae do not appear to have been examined with the electron microscope, but Manton and her colleagues have made an electron microscopic investigation of the structure of spermatozoids of *Fucus*[159] and *Dictyota*[156] and of the zoospore of *Scytosiphon*.[155] From these studies two important features are discussed here: first, the detailed structure of the flagella; second, the specialisation of spermatozoid structure of *Fucus*.

(i) *Structure of the flagella*

The spermatozoid of *Fucus* has two laterally located flagella, the forward projecting one being of the pantonematic type and the posteriorly directed one acronematic. The basal bodies of each are joined laterally, both are in close contact with the nucleus and the front flagellum is attached to the nucleus by fibrous connexions (cf. rhizoplasts of Chrysophyta, Euglenophyta, etc.). The hind flagellum is attached to the cell surface immediately above the eye-spot and the flagellar membrane is dilated at this point. The significance of this in relation to photoreception is unknown. The spermatozoid of *Dictyota* is anomalous in that it has a single flagellum, but examination with the electron microscope reveals two basal granules, the posterior part of the one without a flagellum projecting to the surface of the nucleus.

(ii) *Specialisation of spermatozoid structure*

The spermatozoid of *Fucus* is a specialised structure since the bulk of the cell is occupied by the extremely large nucleus. Other components such as the mitochondria, eye-spot and a vestigial chloroplast are therefore contained within a small part of the cell. In the zoospore of *Scytosiphon*, however, the nucleus does not occupy most of the cell. Instead there is a single large plastid, a prominent eye-spot and many small mitochondria. Apart from the vestigial chloroplast the internal structure of the spermatozoid of *Dictyota* is quite different from that of *Fucus*, and instead it resembles the zoospore of *Scytosiphon*. Since *Dictyota* is generally thought to be a more primitive brown alga than *Fucus* the degree of specialisation of the spermatozoid may be of some phylogenetic significance, although too few observations are, as yet, available.

CLASSIFICATION OF THE PHAEOPHYCEAE

Phaeophyceae can be divided into 11 orders: Ectocarpales, Sphace-lariales, Cutleriales, Tilopteridales, Dictyotales, Chordariales, Sporochnales, Desmarestiales, Dictyosiphonales, Laminariales and Fucales. As for that on the Chlorophyceae the aim of the present chapter is to select from the wealth of detail those funda-mentals of vegetative form, reproduction and life-history which illustrate the essential developments in the class. To do this in so few pages involves a high degree of selection and oversimplifica-tion so that a particular genus or species is described only to illustrate the point under discussion.

Three kinds of life-histories are found in the Phaeophyceae: an isomorphic alternation of stages, a heteromorphic alternation and one in which the only vegetative stage is diploid and in which only the gametes are haploid. These three types have been named Isogeneratae, Heterogeneratae and Cyclosporae respectively.

Forms with an isomorphic alternation of generations (Ectocarpales, Sphacelariales, Cutleriales, Tilopteridales, Dictyotales)

Throughout our discussion of the Phaeophyceae we shall be con-cerned with four features: the vegetative form and its method of growth, the sporangia produced by the sporophyte stage, and the method of sexual fusion.

Species of the Ectocarpales show many of the simplest features in the Phaeophyceae and afford a convenient starting point for a study of these algae. In particular one can begin by enumerating four important features of Ectocarpus siliculosus:

1. The plant is heterotrichous with well-defined prostrate and erect systems.

2. Growth of the erect system is diffuse, but that of the prostrate portion is apical.

3. On the asexual plant, zoospores are formed in both uni-locular and plurilocular sporangia (see p. 23), those from the plurilocular sporangia are diploid and they germinate into the diploid stage whereas those of the unilocular sporangia are hap-loid and they germinate into the sexual phase. The sexual stage produces only gametes formed in plurilocular gametangia.

4. The sexual process is isogamous.

Not all members of the Ectocarpales show the same four features as Ectocarpus siliculosus. Thus, there are five main modi-fications of vegetative form. 1. In several species of Ectocarpus the prostrate system is absent, at least in the mature plant. 2. In others the erect system is absent (e.g. the genus *Streblonema*). 3. Growth is apical in *Ectocarpus lucifugus* and is trichothallic (a well-defined intercalary meristem at the base of a hair) in *E. paradoxus* and *E. irregularis*. 4. In *Pilayella*, cells sometimes divide longitudinally so that a simple parenchymatous form results. 5. The erect threads of some genera (e.g. *Ralfsia*) coalesce to give the plant the appearance of dark brown crusts.

Although most species of Ectocarpales exhibit an isogamous sexual process, *Giffordia secunda* and *Nemoderma* are exceptions, with an anisogamous process. *Nemoderma* (in common with *Pseudolithoderma extensum*) also differs from most other Ectocarpales in that the diploid plant bears unilocular sporangia only. This latter feature is important and is common in the higher orders of Phaeophyceae.

It must be noted that although isomorphic alternation of generations is assumed for all members of the Ectocarpales, details have been established for very few species. This is general throughout the Phaeophyceae and it emphasises the need for continued detailed observation of many different species.

Members of the *Tilopteridales* are also heterotrichous, and the erect system resembles that of the *Ectocarpales*, but cells of the prostrate system are different since they divide in the longitudinal plane so that a simple parenchymatous habit is produced. Unlike the remaining orders in this section, species of Tilopteridales never show apical growth; growth is always intercalary and is usually trichothallic (fig. 24 E).

The order contains only a few genera of which the most common are *Tilopteris* and *Haplospora*, and is recognised by the production of characteristic aplanospores termed *monospores*. These are produced singly in globose sporangia (fig. 24 E). Division is apparently meiotic and, because cleavage of the protoplast does not follow nuclear division, the monospore is quadrinucleate. It is generally assumed that these haploid monospores germinate into a gametophyte, the structure of which is indistinguishable from the sporophyte. Both monosporangia and plurilocular sporangia appear to be borne on the gametophyte whereas monosporangia only are found on the sporophyte.

The *Sphacelariales* show elaboration of several features compared with the two previous orders. Thus, growth is initiated by division of a single apical cell and although division occurs in the transverse plane, the daughter cells also divide longitudinally so that a regular parenchymatous plant body is produced (fig. 24 C).

Vegetative reproduction by the production of *propagules* is common. Such a propagule begins as a normal lateral but vertical division of its apical cell ultimately results in the entire propagule being liberated as a bi- or tri-radiate branch system (fig. 24 D).

The general assumption of an isomorphic alternation in the Sphacelariales is based on the few observations from such species as *Sphacelaria bipinnata, Halopteris scoparia* and *Cladostephus spongiosus*. Diploid plants may (*Sphacelaria bipinnata*) or may not

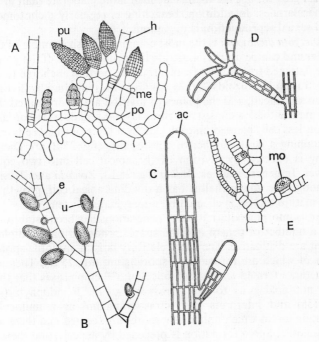

Fig. 24 Phaeophyceae

A, *Ectocarpus confervoides* showing unilocular sporangia. B, *E. reptans* showing plurilocular sporangia. C, D, *Sphacelaria californica*: C, apex of branch; D, mature propagule. E, *Haplospora globosa*. a.c, apical cell; e, erect portion; h, hair; me, meristem at base of hair; mo, monosporangia; po, prostater system; pu, plurilocular sporangia; s, stigma; u, unilocular sporangia. (A, B after Fritsch[91]; C–E after Smith[222])

(*Cladostephus spongiosus*) have plurilocular and unilocular sporangia on the same plant. As in the Ectocarpales, no reduction division precedes spore formation in plurilocular sporangia but does in unilocular. Thus, whereas the diploid zoospores from the plurilocular sporangia perpetuate the sporophyte stage the haploid zoids from unilocular sporangia germinate into an identical gametophytic stage which produces plurilocular sporangia only. Fusion of gametes is isogamous, although physiological anisogamy is known.

Members of the *Dictyotales* are also parenchymatous and show apical growth, but the thallus is much more elaborate than in the Sphacelariales. In addition, branching is regularly dichotomous and sexual reproduction is oogamous.

Dictyota dichotoma is the most common British example of this order and can be used to illustrate the main features of the group. The flattened fronds arise from a rhizome portion anchored to the substratum by rhizoids (fig. 25 A). Growth is initiated by division of an apical cell, and the planes of division are so orientated that the mature thallus consists of three layers; a middle one of large colourless cells banded on each side by a layer of small cells each containing a large number of chloroplasts (fig. 25 B–D). Dichotomy is effected by division of the apical cell into two equal halves. In genera such as *Padina* (fig. 25 E), *Zonaria* and *Taonia*, although the young thallus has a dividing apical cell, growth of the mature thallus is effected by a marginal meristem.

An isomorphic alternation of generations has been established for a number of genera and the spores produced on the diploid plant are distinctive. They are relatively large non-motile spores, four of which are formed in a sporangium (fig. 25 D). They are sometimes termed *tetraspores*, but Smith[222] emphasises that they are not analagous to the tetraspores of the Rhodophyta (see p. 156) and interprets the 'tetrasporangium' as a unilocular sporangium in which only a few spores are formed and these are non-motile. Spore formation is preceded by meiosis, and there is genotypic determination of sex at division, so that two of the spores germinate into the female gametophyte and two into the male. Both antheridia and oogonia (on separate plants) develop from surface cells and are borne in sori on both sides of the thallus. The superficial cells developing into oogonia elongate, and the ovum is formed singly in the oogonium (fig. 25 B). The antherideal initial divides a number of times to give a large number of compartments the contents of each of which develop

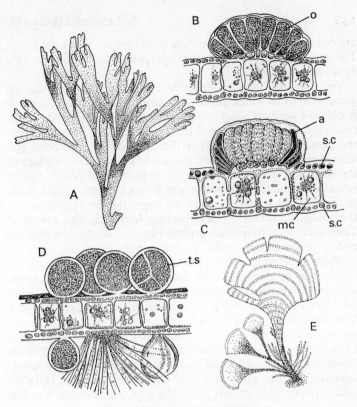

Fig. 25 Phaeophyceae: Dictyotales

A-D, *Dictyota dichotoma*: A, habit; B, cross section through an
oogonial sorus; C, cross section through an antheridial sorus;
D, tetrasporangia. E, *Padina pavonia*, habit. a, antheridium (a
large number of small antheridia in each sorus); m.c, large median
cell of thallus; o, oogonia, 20-50 in each sorus, each with single egg;
s.c, surface layer of small cells; t.s, tetraspores in tetrasporangium.
(From Fott[84])

into a single uniflagellate antherozoid (fig. 25 C). The eggs are
liberated by gelatinisation of the oogonial wall and fertilisation
follows immediately.

The thallus of *Cutleriales* is also parenchymatous, but unlike
the Sphacelariales and Dictyotales growth is not apical but occurs
by a specialised trichothallic process. The meristematic cells at
the base of hairs divide to give outer daughter cells which contribute to the length of the hair and inner ones which subsequently

divide longitudinally to contribute to the main part of the frond. The mature thallus consists of inner large cells bounded by smaller epidermal cells (fig. 26 B).

The order contains two genera, *Cutleria* and *Zanardinia*, the gametophytes of which are superficially very different. *Zanardinia* is disc-like whereas most species of *Cutleria* are irregular, dichotomously branched, flattened thallus (fig. 26 A). In both genera the gametangia develop either directly from a superficial epidermal cell or from uniseriate threads arising from the surface of the thallus. The initials divide into a number of compartments (usually a smaller number in the female). Each compartment produces a single swarmer and the larger female ones soon lose their flagella, round off and become fertilised. The zygote of *Zanardinia* germinates into a plant identical to the gametophyte and it produces unilocular sporangia only. *Cutleria*, however, exhibits a heteromorphic alternation of generations since the sporophyte is disc-like. The disc-like sporophyte used to be assigned to a separate genus, *Aglaozonia*, until culture work established its relationship to *Cutleria*. Although *Cutleria* has a heteromorphic alternation of generations it is placed in the present section because of its obvious affinities to *Zanardinia*. Fritsch[91] has made the interesting suggestion that *Cutleria* is the remaining erect system and *Aglaozonia* the remaining prostrate system of a plant which was once heterotrichous. That is, in the gametophyte stage of this plant the prostrate system was lost whereas in the sporophyte the erect system disappeared.

Forms with a heteromorphic alternation of generations (*Chordariales, Sporochnales, Desmarestiales, Dictyosiphonales and Laminariales*)

Members of this group show a regular alternation between a diploid sporophyte and a morphologically different haploid gametophyte stage. The thallus is either pseudoparenchymatous (*haplostichous*) formed by aggregation of filaments, or parenchymatous (*polystichous*) formed by early longitudinal division of the cells.

Haplostichous forms (*Chordariales, Sporochnales, Desmarestiales*)

Members of the *Chordariales* show the simplest kinds of pseudoparenchymatous thalli and many are small microscopic filamentous-like bodies. Thus, the sporophyte of *Myrionena*, for example, consists of radiately branched horizontal filaments which tend to

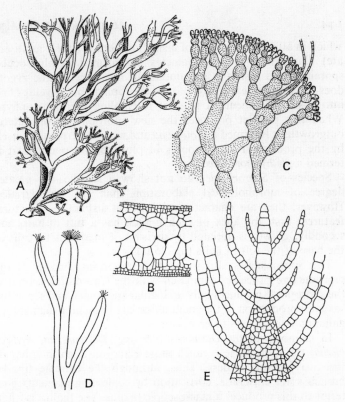

Fig. 26 Phaeophyceae:
Cutleriales, Chordariales, Sporochnales and Desmarestiales

A, gametophyte of *Cutleria multifida*, habit. B, transverse sections
of gametophyte of *C. adspersa*. C, *Leathesia difformis*, vertical
section through outer portion of sporophyte. D, *Carpomitra
cabrerae*, habit. E, *Desmarestia herbacea*, surface view of growing
apex. (After Smith[222])

coalesce as a flat disc and from which, on the upper surface, a
number of erect filaments develop. *Leathesia* is more elaborate
and consists of a hollow much convoluted globose thallus, the
solid part of which consists of radiating filaments (fig. 26 C).

Both unilocular and plurilocular sporangia, plurilocular alone
or unilocular alone can be borne on the sporophyte. The zoospores produced in the plurilocular sporangia are diploid and they
germinate into the sporophyte stage. The haploid zoospores from
the unilocular sporangia germinate into gametophyte stages

which in all genera consists of a small uniseriate (rarely multiseriate) filamentous thallus. Gametes are formed in plurilocular sporangia and fusion is isogamous. In most species the zygote does not germinate directly into the sporophyte but develops first into a small filamentous plantlet resembling the gametophyte. When this plantlet produces the new sporophyte from lateral outgrowths it is termed a *protonema* and when it perpetuates itself by the production of zoospores in plurilocular sporangia it is termed a *plethysmothallus*.

Species of *Sporochnales* do not show a significantly greater degree of morphological elaboration than the Chordariales. However, they are characterised by two distinctive structural features: first, the apex of each branch has a tuft of hairs and secondly, growth is initiated by division of meristematic cells at the base of the tuft of hairs (fig. 26 D).

The Sporochnales also shows two important changes of the reproductive processes compared with the Chordariales. Firstly, the sporophyte produces only unilocular sporangia, and secondly, sexual reproduction of the small unisariate branching filamentous gametophyte appears to be oogamous.

In members of the *Desmarestiales* (e.g. *Desmarestia*, *Arthrocladia*) the sporophyte is much more elaborate than in either of the two previous orders. Thus, although the extreme tips of branches are uniseriate, cortication by coalescing filaments posterior to this produces a macroscopic thallus. The thallus is often flattened and is always pinnately branched. The apex of each branch is terminated by a hair and growth is trichothallic with a well-defined zone of meristematic cells at the base of the hair. Division of these cells has three effects: first, to increase the length of the hair above; secondly, to add to the axial filament below; and thirdly, to produce uniseriate lateral hair-like branches. It is division of cells at the base of these hair-like laterals which produce the corticating filaments (fig. 26 E).

The sporophytes of both *Desmarestia* and *Arthrocladia* produce only unilocular sporangia and the haploid zoospores derived therefrom germinate into small uniseriate gametophytes. *Desmarestia* is dioecious and sexual reproduction is oogamous. Antheridia are borne in clusters at the tips of branches and oogonia are formed by elongation of some intercalary cells. The single large ovum is extruded from the apex of the oogonium and is fertilised by the antherozoid while still attached to the apex of the oogonial wall.

Polystichous forms (Punctariales, Dictyosiphonales, Laminariales)

The gametophytes of all three orders are sparsely branched microscopic thalli, whereas the sporophytes are parenchymatous of varying degrees of elaboration.

The form of the sporophyte in the *Dictyosiphonales* is variable. For example, *Stictyosiphon* consists of branching endophytic threads, the cells of which occasionally divide longitudinally. *Punctaria* is more elaborate and consists of leaf-like thalli closely resembling *Ulva* except that the thallus is 7-celled thick with the inner cells frequently longer than the outer. Anatomical differentiation also occurs in *Soranthera* which is a hollow subspherical plant with the cellular portion consisting of inner large cells with an outer epidermis-like layer.

Accompanying this increasing morphological elaboration is an increasing tendency for the sporophyte to produce only unilocular sporangia. Thus, this is so for *Soranthera* whereas the sporophytes *Stictyosiphon* and *Punctaria* bear both unilocular and plurilocular. Growth of the thallus is diffuse and sexual reproduction is usually anisogamous.

Other members of the *Dictyosiphonales* (for example, *Dictyosiphon*) are more elaborate, consisting of branched cylindrical thalli, the mature parts of which are differentiated into an inner zone of large cells with outer epidermal layers of small cells. Growth is apical, and sporophytes produce unilocular sporangia only; sexual reproduction is isogamous.

Species of *Laminariales* also exhibit alternation between a complex macroscopic sporophyte and a microscopic filamentous gametophyte, but the size and degree of anatomical differentiation of the sporophyte is greater than is found elsewhere in the algae. The thallus of most species (the simple, whip-like *Chorda* is an exception) is differentiated into a holdfast, stipe and lamina and in most species, too, growth is initiated by division of meristematic cells located at the transition between stipe and lamina. Division of these cells contributes to new laminar tissue above and to new stipe material below. Both tissues undergo considerable differentiation from an early stage and the anatomy of the mature stipe resembles that of the mature lamina except for the displacement due to flattening of the blade. One of the earliest processes in differentiation is tangential division which produces a superficial layer separated from an inner group of cells. The

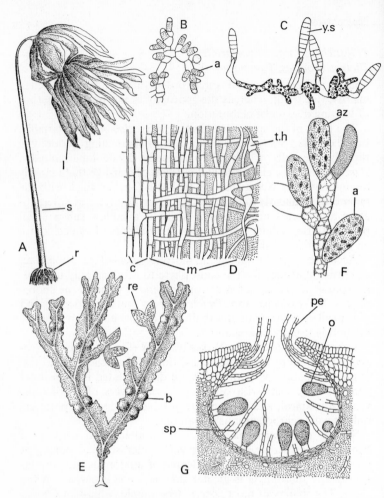

Fig. 27 Phaeophyceae: Laminariales and Fucales

A, *Laminaria hyperborea*, habit. B, C, *L. flexicaulis*: B, male gametophyte with antheridia; C, female gametophyte with young sporophytes. D, *L. andersonii*, diagrammatic longitudinal section through medulla of stipe. E, *Fucus vesiculosus*, habit. F, *Fucus furcatus*, antheridial paraphysis. G, *F. vesiculosus*, section through oogonial conceptacle. a, antheridia; a.z, antherozoids; b, air bladders; c, cortical region; l, lamina; m, medulla; o, oogonium; pe, periphyses; r, rhizoid; re, receptacle; s, stipe; s.p, sterile paraphyses; t.h, 'trumpet' hyphae; y.s, young sporophyte. (A after Fritsch[91]; B-G after Smith[222])

latter do not divide further and constitute the medulla of the mature plant, whereas the superficial layer functions as a meristem, the inner daughter cells forming a cortex and the other daughter cells forming a superficial 'meristoderm' layer. The structure of the medulla is further elaborated by cross-connecting 'hyphae', etc., which are produced during maturation (fig. 27 D).

The sporophytes produce only unilocular sporangia which are borne, together with sterile paraphyses, in sori. They usually occur on the normal laminae of the plants but in *Alaria* and related genera they are produced on specialised sporophylls. Division in the sporangia is meiotic and the haploid zoospores germinate into a microscopic filamentous plantlet (fig. 27 B, C). The gametophytes are dioecious and sexual reproduction is oogamous and the ovum is usually fertilised while attached to the oogonial wall.

The tendency for the sporophyte to become increasingly elaborate whereas the gametophyte becomes more reduced culminates in the *Fucales*, in which there is no gametophyte stage.

As in the Laminariales the sporophyte of the *Fucales* is morphologically complex (e.g. fig. 27 E) with a high degree of anatomical differentiation. Unlike the Laminariales, however, growth of the Fucales is by a small apical cell although there are a few exceptions (e.g. *Hormosira* has several apical cells, and the growth of *Durvillea* is diffuse).

The Fucales do not produce spores and reproduce solely by an oogamous sexual process. The oogonia and antheridia are located in cavities on the blade (*conceptacles*) consisting essentially of a lining layer of cells derived by vertical division of an initial superficial cell. The upper part of the lining layer produces hairs which project to the outside (through the *ostiole*) and the basal part produce paraphyses on which the antheridia are normally borne (fig. 27 F). The oogonia, on the other hand, are borne directly on the conceptacle wall (fig. 27 G). Sixty-four biflagellate, pear-shaped antherozoids are normally produced, but the number of eggs is more variable. In all genera, nuclear division produces 8 haploid nuclei and in *Fucus* the eight persist whereas in others 4, 6 or 7 of them degenerate. Fusion takes place in the water after the eggs are released from the oogonia.

RHODOPHYTA: RHODOPHYCEAE

Division RHODOPHYTA
Class RHODOPHYCEAE

Species of Rhodophyceae are characterised by five main features. 1. There are no flagellated stages. 2. Their photosynthetic pigments include the characteristic biloproteins R-phycocyanin and R-phycoerythrin, together with chlorophyll d and the carotenoid taraxanthin. 3. Food storage products include floridean starch and the galactoside floridoside. 4. Polysulphate esters are characteristic components of the cell walls. 5. They exhibit a highly specialised and characteristic process of sexual reproduction.

CELL STRUCTURE

A *cell wall* is always present, and usually contains cellulose and pectic material together with polysulphate esters. X-ray analyses and electron microscopic investigations[175, 176, 177] have revealed a similar wall structure in a number of different species, namely that of cellulose microfibrils scattered through a granular matrix. The microfibrillar fraction is significantly lower than in other algae.

Cell walls of some red algae have *pits*, and these appear to allow protoplasmic continuity between adjacent cells (there is no general agreement on this). Their structure is variously described as open, perforated or enclosed by a membrane. Examination with the electron microscope would be expected to clarify much of the confusion but as yet it has not done this (cf. refs. 13 and 177).

Cells of the Bangiophycidae normally contain a single, more or less stellate axile chloroplast, whereas those of the Florideophycidae normally have a number of parietal chloroplasts. This difference is not absolute, however, and there is much variability. More heavily stained areas are chiefly found in axile chloroplasts (although not exclusively so) and although these have been termed pyrenoids their functional equivalence to the pyrenoids of Chlorophyceae has not been established. In fact, grains of Floridean starch are usually deposited outside the chloroplasts and the smaller grains are frequently associated with the cisternae of the endoplasmic reticulum[13] (cf. the deposition of animal glycogen).[200]

Although the gross form of chloroplasts is variable their ultrastructure shows remarkable uniformity. Thus, electron microscopic investigations of many different species, from both the Bangiophycidae and the Florideophycidae, have confirmed that the membrane-bound chloroplast has a homogeneous matrix traversed by a number of bands, and in all cases each band has a *single disc*. This fact contrasts with that found in most algae and may be of phylogenetic importance.

Cells of the Bangiophycidae and of the Nemalionales, Cryptonemiales and Gigartinales of the Florideophycidae usually have a single nucleus whereas in other orders the cells tend to be multinucleate. The nucleus has a well-defined membrane, usually a prominent nucleolus (sometimes more than one) and a mitotic division process has been observed.

Cells normally have one or more vacuoles and those of *Lomentaria* are bounded by a tonoplast membrane. Electron microscopic investigations have also confirmed that cells of *Lomentaria*[13] and *Porphyridium*[17] have dictyosomes and an endoplasmic reticulum. Although mitochondria have been identified in *Lomentaria* none have yet been seen in *Porphyridium*.

CLASSIFICATION OF RHODOPHYCEAE INTO BANGIOPHYCIDAE AND FLORIDEOPHYCIDAE

The fundamental separation of the Rhodophyceae into two subclasses, the Bangiophycidae and Florideophycidae, has been emphasised for many years and has been based on a number of features. 1. Species of Bangiophycidae are unicellular, simple or branched filaments, solid cylinders or flattened sheets of cells, but

never show aggregation of filaments into a pseudoparenchymatous structure as is characteristic of the Florideophycidae.
2. In the Bangiophycidae, growth of the thallus is diffuse, whereas
it is usually apical in the Florideophycidae. 3. Cells of the Bangiophycidae normally have a single axile stellate chloroplast whereas
a number of parietal chloroplasts is most common in the Florideophycidae. 4. In the Bangiophycidae the production of carpospores is by direct division of the zygote whereas in the Florideophycidae they are formed indirectly from the zygote (see pp. 151
and 156). 5. Pit connections are more common in the Florideophycidae than they are in the Bangiophycidae.

The presence or absence of pit connections used, at one time,
to be considered a character of fundamental difference between
the two subclasses. However, there have been an increasing
number of observations of such structures in the Bangiophycidae,[77] and Dixon[49] has recently analysed the taxonomic significance of these observations. He points out that, although their
presence or absence can no longer be accepted as an absolute
difference between the two subclasses, pit connections are more
common and prominent in the Florideophycidae.

BANGIOPHYCIDAE

VEGETATIVE STRUCTURE

Although this subclass contains only about 15 genera their
vegetative form is extremely variable, ranging from unicellular
forms (e.g. *Porphyridium*) to filamentous (e.g. *Goniotrichum*,
fig. 28 A, B) and parenchymatous forms (e.g. *Porphyra*, fig. 28 C).
Classification of the Bangiophycidae is based primarily on the
vegetative structure, and the orders Porphyridiales, Gonio-
trichales and Bangiales have been suggested[191] to accommodate
the unicellular, filamentous and parenchymatous forms respectively. Alternative classification schemes suggest three families
within the single order, the Bangiales.

Although the membranous expanses of *Porphyra* can reach a
considerable size, the basic structure of all parenchymatous forms
is simple. In the early stages the thallus appears as a uniseriate
filament, the cells of which subsequently divide longitudinally.
Growth of all multicellular forms is diffuse.

The form of chloroplast in the Bangiophycidae is more variable
than in the Florideophycidae. For example, *Porphyridium* has a

single axile stellate chloroplast with a pyrenoid, whereas the filamentous form *Goniotrichopsis sublittoralis* has several parietal
disc-shaped chloroplasts without a pyrenoid.

ASEXUAL REPRODUCTION

The most common type of multiplication of *unicellular forms* is
vegetative division of the cell into two daughter cells. However,
division of the cell contents into more than one daughter cell has
been described for *Porphyridium*. Although Geitler[94] has observed
a budding process in *Porphyridium cruentum*, he considers this to
be a pathological process which does not occur under normal
conditions.

Multicellular forms produce various kinds of spores.[63] Mono-
spores are produced either in cells indistinguishable from vegetative cells (e.g. *Porphyra*, *Bangia*, *Goniotrichum*), or in sporangia
which are smaller than the vegetative cell (e.g. *Erythrotrichia*,
Rhodochaete, *Kyliniella*). When the cell contents divide to produce
more than one spore, the spores are generally interpreted as being
carpospores (fig. 28 D) and so are discussed in the following
section.

SEXUAL REPRODUCTION

The male sex organ is a *spermatangium* and it liberates a single
non-motile *spermatium*. The female sex organ is the *carpogonium*,
and is unicellular with the distal end prolonged as a trichogyne.

As was stated above, the spores produced by successive
divisions of cell contents are generally interpreted as carpospores.
That is the mother cell is thought to be a zygote (this direct
formation of carpospores by division of the zygote contrasts with
the indirect formation of carpospores in the Florideophycidae).
Actual fusion of male and female gametes has been observed in
such genera as *Porphyra*, *Bangia*, etc. (Drew[63] discusses the
evidence in detail).

Life-histories of Bangia and Porphyra

The carpospores of both *Bangia* and *Porphyra* germinate into a
minute filamentous stage resembling *Conchocelis rosea* (fig. 28 E).
It is generally thought that germination of the zygote is accompanied by meiotic division of the nucleus, and that the life-history
of these forms involves an alternation between a diploid (*Porphyra*, *Bangia*) stage and a haploid stage (*Conchocelis*). However,
the evidence is not conclusive, and alternative suggestions have

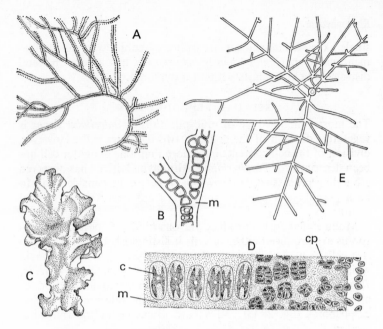

Fig. 28 Rhodophyta: Bangiophycidae

A, B, *Goniotrichum elegans*: A, habit; B, part of the branching threads. C, D, *Porphyra perforata*: C, habit; D, vertical section of thallus showing vegetative cells, carposporangia and carpospores. E, *Conchocelis* phase. c, chloroplast; cp, carpospores (remains thereof in E); m, mucilage. (A, B after Fritsch[91]; C, D after Smith[222]; E after Drew[61])

been made. For example, Dangeard[38] interprets the *Conchocelis* stage as a protonemal phase. A particularly important investigation of Magne[152] casts doubt on the generally accepted picture, since he observed that the germination of the zygote of *Rhodochaete parvula* was not accompanied by meiotic division of the nucleus.

FLORIDEOPHYCIDAE

VEGETATIVE STRUCTURE

Despite the apparent great variability of vegetative form in this subclass, there are a number of basic features, and Drew[60] has summarised these as follows:

1. All thalli are fundamentally filamentous. That is, even the more elaborate forms are not truly parenchymatous, but are

pseudoparenchymatous, consisting of coalescing filaments. Their fundamental filamentous nature is frequently difficult to see, except in the younger portions.

2. Growth of all filaments is generally apical, and there is a basic pattern of division. Cells formed by division of the apical cell divide tangentially to produce pericentral cells outside the main axis. The further development of these cells determines the degree and type of elaboration. In the simplest Rhodomelaceae, for example, the pericentral cells remain unchanged; in others they divide transversely and so constitute a cortex around the central row of cells, and in others they function as initials of branches. Growth and development of laterals is also important in determining the final form, and there are two common kinds of development; firstly, division of the basal cells of the branch to form corticating threads, and secondly, repeated branching of the laterals until the terminal ones unite sharing a compact structure with a medulla and a cortex.

3. Two main types of construction have been recognised: *uniaxial* and *multiaxial*; the former having a single main filament and the latter with several axial filaments.

Uniaxial forms

Monosiphonous filaments can be found in genera such as *Rhodochorton* and *Antithamnion*, while the genus *Batrachospermum* shows the simplest kind of elaboration in which three or four branches arise in whorls and cortication of the central filament is effected by down-growing filaments developed from the basal cells of those branches (fig. 29 A). Genera such as *Calosiphonia* and *Dumontia* are more elaborate due to confluent growth of the branches producing a continuous outer layer, the cortex (fig. 29 B). Yet more elaborate forms such as *Plocamium*, *Ceramium* (fig. 29 C), *Polysiphonia* and *Hypoglossum* are described by Drew,[60] who provides an excellent summary of the many kinds of elaborations found amongst uniaxial forms (ref. 60, p. 174).

Multiaxial forms

Drew[60] also outlines the essential features of those thalli with more than one axial filament. The increasing elaboration shown by such genera as *Helminthocladia*, *Galaxaura*, *Furcellaria* and *Chylocladia* is comparable to that outlined for the uniaxial forms. Thus, *Helminthocladia* has several axial filaments each with an apical cell and the laterals produced from these central filaments

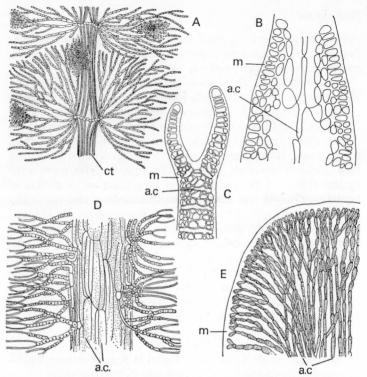

Fig. 29 Rhodophyta: structure of Florideophycidae

A, *Batrachospermum moniliforme*. B, *Dumontia incrassata*. C, *Ceramium deslongchampsii*, apex of branch. D, *Helminthocladia divaricata*. E, *Furcellaria fastigiata*; a.c, axial cells (A-C, uniaxial, D, E, multiaxial); c.t, corticating threads obscuring single axial cell; m, mucilage. (All after Fritsch[91])

remain distinct (fig. 29 D). In *Galaxaura*, however, the laterals coalesce to give a more or less compact cortex with a tendency for downgrowing hyphae to obscure the main axis. *Furcellaria* shows even greater structural elaboration with small-celled outer cortex, an inner cortex with large cells and a medulla with axial filaments and secondary hyphae. In *Chylocladia*, also, coalescing branches form a compact cortex but in this genus the axial filaments separate and surround a hollow, central cylinder.

REPRODUCTION

Apart from asexual reproduction by the production of mono-spores the most common method of reproduction in the Florideophycidae is a characteristic and elaborate sexual process. The details of this process are variable and are important criteria for the major classification of the subclass, and these numerous variations are presented later in the chapter when discussing classification of the Florideophycidae. For the moment the basic outline of the process will be presented (see diagrammatic representation in fig. 30).

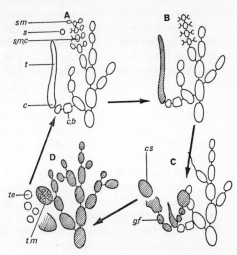

Fig. 30 Rhodophyta: reproduction of Florideophycidae

Diagram of triphasic life history in the Florideophycidae. A, gametophyte with spermatum about to fertilise a carpogonium. B, gametophyte with diploid carpogonium. C, diploid carposporophyte on the gametophyte. D, free-living diploid tetrasporophyte. c, carpogonium; c.b, 'carpogonial branch'; cs, carpospore; g.c, haploid cell of gametophyte; g.f, gonimoblast filament of carposporophyte; s, spermatium; sm, spermatangium; s.m.c, spermatangial mother cell; t, trichogyne; te, tetraspore; tm, tetrasporangium. (After Smith[222]) (Diploid cells are shaded)

The male gametes (*spermatia*) are non-motile and are formed in *spermatangia*. These latter structures are usually formed in large numbers in close proximity to one another, and in more specialised species are localised in well-defined sori (e.g. *Apoglossum.* The spermatia are borne passively to the female organ, the *carpogonium*, which consists essentially of a basal egg cell and a protuberance of variable length and shape (the *trichogyne*). The carpogonia are normally produced (in smaller numbers than the spermatangia) as terminal cells of an accessory lateral, the *carpogonial branch* (for a discussion of the concept of a carpogonial branch, see Dixon[48]). The spermatia become attached to the trichogyne and the male nucleus passes down to fuse with the female nucleus at the base of the egg cell.

Apart from certain details, the process described up to this point is relatively uniform for all Florideophycidae; the variability is introduced when we consider the germination of the zygote. It is clearly not possible to present details of the variations and complexities of the process in a small book. Instead, the essential features which form the basis of the process in all species will be presented. Also, the process will be described in such a way as to suggest a firm, well-established sequence of events. However, this is not so, and much of the inference and generalisation rests on scanty direct evidence obtained from only a few species. The uncertainties will be mentioned at the end of the chapter when inconsistencies and errors in classification are described.

Unlike the Bangiophycidae, carpospores are not produced directly by division of the zygote. Instead they are borne on a distinct stage, the *carposporophyte*, which develops as a filamentous outgrowth (the *gonimoblast filaments*) from the gametophyte (fig. 30 C). Nuclei of the carposporophyte are derived ultimately from the zygote nucleus and in most species nuclear division is mitotic so that the carposporophyte is diploid. The carposporophyte reproduces by the production of carpospores. In the few forms with haploid carposporophytes the spores germinate into the gametophyte stage, so that there is an alternation between two morphologically different phases, both haploid, one (the gametophyte) reproducing sexually and the other (the carposporophyte) reproducing by carpospores. In most species, however, the diploid carpospores produced by the diploid carposporophyte do not germinate into the gametophyte and instead they develop into the *tetrasporophyte* stage. This is morphologically identical to the gametophyte, but instead of producing spermatia and carpogonia it reproduces by the production of *tetraspores*. The four non-motile spores are produced in *tetrasporangia* (fig. 30 D) and, depending on the method of division, different kinds of sporangia can be recognised. In *cruciate* sporangia the first division is transverse and the second is longitudinal (one in each of the two daughter cells, usually at right angles to one another). *Zonate* sporangia have three parallel transverse walls and *tetrahedral* forms have the spores arranged tetrahedrally. Nuclear division during tetraspore formation is meiotic and the haploid tetraspores germinate into the gametophyte.

Although the nuclei in the carposporophyte are derived ultimately from the zygote nucleus the gonimoblast filaments do not always arise from the zygote itself. In many species the zygote nucleus migrates to a specialised vegetative cell, the *auxiliary cell*. The position of the auxiliary cell is variable and can be a relatively great distance from the zygote so that the nucleus has to migrate in a well-defined tube.

CLASSIFICATION OF THE FLORIDEOPHYCIDAE

The most widely accepted system of classification is that which divides the Florideophycidae into 6 orders: Nemalionales, Gelidiales, Cryptonemiales, Gigartinales, Rhodymeniales and Ceramiales. This system is based largely on early investigations by Schmitz[218] and the later refinements of Kylin and his co-workers.[136] Although it is generally presented as a firm, well-established system, Dixon[46] has pointed out that 'the apparent stability of the system of classification is largely an illusion'. However, no alternative well-supported classification has yet been presented. Because of this, the discussion below is in two parts: the first outlining the generally accepted classification and the second presenting some of the more serious objections to this. Thus many of the statements in the first part will be refuted later.

1. Most of the orders have two diploid stages (the carposporophyte and the tetrasporophyte) alternating with a haploid gametophyte stage. The *Nemalionales* is generally separated as one in which the zygote is thought to germinate into a haploid carposporophyte.

2. The remaining five orders can then be divided on the basis of the presence or absence of auxiliary cells; these are usually thought to be absent from the *Gelidiales* but present in all others.

3. Of the remaining four orders, *Cryptonemiales* and *Gigartinales* are characterised by the fact that their auxiliary cells differentiate before sexual fusion, whereas in the *Rhodymeniales* and *Ceramiales* they differentiate after fusion.

Further details of each order are as follows:

1. *Nemalionales*

Division of the zygote nucleus is usually meiotic except in *Nemalion*, *Scinaia*, *Lemanea* and members of the Bonnemaisonaceae[153] (further details of exceptions are given on p. 159). In most species of the order the carposporophyte develops from the carpogonium, but in a few the zygote nucleus passes to a vegetative cell of the carpogonial filament, and there has been much controversy as to whether this cell receiving the zygote nucleus is a 'typical' auxiliary cell (cf. refs. 135 and 236). The order contains some of the simplest types of Florideophycidae such as *Rhodochorton*, but also has more elaborate forms such as *Atractophora* (uniaxial) and *Scinaia* (multiaxial).

2. *Gelidiales*

This order is characterised by the absence of auxiliary cells. Also the carpogonial filaments are one-celled, and the cell metamorphoses into a carpogonium. Tetrasporangia are cruciate. The order contains only a few genera, all of which have a uniaxial type of construction (e.g. *Gelidium, Pterocladia*).

3. *Cryptonemiales*

The auxiliary cells differentiate before fusion, and are borne in special accessory branches. These may be near or remote from the carpogonial branches; the latter also arise on special accessory filaments and in the Corallinaceae (e.g. *Corallina*) they are located in well-defined *nemathecia* (sori). After fertilisation, tubular outgrowths (ooblasts) are formed connecting the carpogonium with the auxiliary cell, and the zygote nucleus passes along it. Tetrasporangia are either cruciate or zonate. Species are either uniaxial (e.g. *Dudresnaya, Cryptosiphonia*) or multiaxial (e.g. *Corallina*).

4. *Gigartinales*

In this order, also, the auxiliary cells differentiate before fusion, but unlike the previous one the carpogonial filaments arise from the normal vegetative thallus and not from specialised accessory filaments. Moreover, unlike the situation in other orders the auxiliary cells are intercalary cells of the normal vegetative thallus, or in some species are supporting cells of the carpogonial filament. Thalli may be based either on the uniaxial type of construction (e.g. *Sphaerococcus*) or multiaxial (e.g. *Gigartina, Phyllophora*).

5. *Rhodymeniales*

Although the auxiliary cell does not differentiate as such until after fertilisation, the small immature auxiliary cell which is indistinguishable from a vegetative cell is formed before fusion. Tetrasporangia are cruciate or tetrahedral. All species are multi-axial and in many genera (e.g. *Champia, Chylocladia, Lomentaria*) the mature thallus has a hollow centre divided at intervals by septa, or diaphragms.

6. *Ceramiales*

In this order the auxiliary cell is borne directly on the supporting cell of the carpogonial filament and not only does it not differentiate until after fertilisation but unlike that of the previous order it is not cut off until after fusion. The carpogonial filament is always four-celled and the tetrasporangia are usually tetrahedral but are sometimes cruciate. The vegetative form of this order is extremely variable. Some genera are branched filaments (*Callithamnion*) and others show various types of pseudoparenchymatous thalli ranging from leafy expanses (*Delesseria*), cartilaginous forms (*Laurencia*) to various polysiphonous forms (*Polysiphonia*). All species of the pseudoparenchymatous forms are of the uniaxial type.

The extensive variability and diversity of reproduction and vegetative structure within each of the six orders cannot be indicated by the summary above and the larger texts and monographs[84, 91, 136] are recommended for these numerous important details.

Although the classification has been presented as a firm, well-established system, several recent publications have indicated that it is sometimes based on doubtful evidence and that a more critical examination of the data might result in extensive modifications of the generally accepted classification. Some of the more important doubts are summarised as follows:

1. *Life-histories in the Nemalionales and Gelidiales*

It is generally assumed that there is no tetrasporophyte stage in the Nemalionales and that division of the zygote nucleus is meiotic. However, this is an example of a generalisation based on few observations and is, in fact, based almost entirely on Svedelius'[236] observation of meiotic division of the zygote of *Scinaia*. Dixon[46] has recently discussed the evidence for tetrasporophytic stages in a number of different genera (e.g. *Actinotrichia, Delisia pulchra* and *Leptophyllis conferta*, and *Rhodochorton violaceum*). He concludes that, although the earliest evidence is inconclusive and confusing, more recent observations make the hypothesis that a tetrasporophytic stage is absent from the Nemalionales untenable.

Generalisations of the life-histories in the Gelidiales are also based on too few observations. Thus, the alternation of two diploid and one haploid stage is assumed solely from observations of the existence of sexual and tetrasporophytic stages and Drew[62] has shown how this conclusion is based on scanty direct evidence.

2. *The concept of the 'auxiliary cell' in the Gelidiales*

It is generally assumed that members of the Gelidiales do not have a 'typical' auxiliary cell. This conclusion is supported by Kylin who defines an auxiliary cell so as to exclude the cell in the Gelidiales. However, there are many authorities who disagree with this interpretation. Dixon[46] has summarised these latter opinions and it appears that, unless a highly suspect definition is accepted, the absence of a typical auxiliary cell can no longer be accepted.

3. *Relationship between the Gelidiales and Nemalionales*

The distinction between these two orders rests on their life-histories; the two haploid stages of Nemalionales supposedly contrasting with the alternation between diploid and haploid in the Gelidiales. Since this difference has been shown to be doubtful (section 1 above) Dixon[46] has presented his reasonings for including the Gelidiaceae as a separate family within the Nemalionales. This makes the Nemalionales a very heterogeneous order but, as Dixon points out, comparable diversity remains even when the Gelidiaceae are removed.

4. *Relationship between the Cryptonemiales and the Gigartinales*

Cryptonemiales are usually distinguished from the Gigartinales by the position of the auxiliary cells; in the former they occur on special accessory branches while in the latter they occur as intercalary cells of a normal vegetative lateral. Fritsch considers that it is rather a trivial feature upon which to base a major taxonomic sub-division (ref. 91, p. 656). He further emphasises the parallel and overlapping types; although, as Papenfuss remarks, 'whether the types represent a true relationship or merely instances of parallel evolution cannot at present be determined' (ref. 186, p. 10). Although Drew[62] also recognised the overlapping types, she emphasised that no evidence was available on which to base a major regrouping of families of the two orders.

5. *Auxiliary cells in the Ceramiales*

The formation of an auxiliary cell from the supporting cell after fertilisation has been described above as a diagnostic feature of the Ceramiales. However, Dixon[54] has recently discussed the evidence for the supporting cell functioning directly as the auxiliary cell. He has observed such a phenomenon in *Griffithsia globulifera* and *Antithamnion spirographidis*. He points out the inadequacy of Hommersand's suggestion that such phenomena are abnormalities arising from the close proximity of other 'normal' carposporophytes.

6. *Uncertainties at the species and generic level*

The examples of uncertain or erroneous classification enumerated above affect the creation of major taxonomic groups. As would be expected, therefore, more detailed and critical observation also reveals many errors and uncertainties in the classification of individual genera and species of the Florideophycidae. An appreciation of the details demands a background knowledge of genus and species which has not been presented in this book, so that this present section aims only to mention some of the relevant literature. Fortunately, Dixon, who has encountered many of the problems when preparing the volume on *Rhodophyta* for the proposed *Flora of British Marine Algae*, has published his observations and comments.[44, 45, 47, 50, 52, 53] Although no details can be given, the extent of the uncertainties can be illustrated by citing Dixon, who, when describing the difficulties of defining species in the Rhodophyta, predicts that 'about 20% of the species listed in the current check-list of British marine algae will be eliminated eventually and it is doubtful if more than about 70% can be identified accurately in any flora or taxonomic treatise, for any part of the world' (ref. 50, p. 52). Clearly, therefore, a major reconstruction of the classification of the Rhodophyta is needed, and until one is proposed, all that a small book can hope to do is to present the fundamentals of the most widely accepted classification and to present a summary of recent objections to this.

PHYLOGENY OF THE ALGAE

A phylogenetic classification of any group of organisms is one which attempts to illustrate the evolutionary relationships between members of the group. An alternative system is one whose sole (or at least chief) aim is to allow clear and well-defined limits to be drawn around any species, or group of species. The two systems are not of course mutually exclusive, since it is probable that a scheme based on evolutionary relationships is also more likely to provide a stable utilitarian system.

Although the accuracy of any phylogenetic classification cannot be assessed by applying any single rule, one important requirement is that groupings shall be based on several (rather than one or two) features. This does not imply that classifications based on a single criterion are not useful; but it does mean that confirmation with other criteria is needed to minimise any tendency towards artificial groupings.

With reference to algae, the fact that a classification into the major phyla made on the basis of photosynthetic pigments is supported by other criteria such as storage products, cell wall constituents, nature of flagella and details of cell structure suggests that the classification is natural. That is, that divergence into distinct phyla probably occurred relatively early in the evolution of algae. Because the classification into phyla has a phylogenetic basis, it can be used as a basis of phylogenetic discussions, and within this context the discussions have two aspects: first, evolutionary trends within each phylum, and secondly, relationships between different phyla.

Any discussion of possible evolutionary relationships is frequently speculative; this is particularly so for discussions of algal evolution, because fossil evidence is frequently ignored or at least minimised. Observations of fossil diatoms,[202, 220] dinoflagellates,[75, 213, 239] members of the Dasycladales[198] and a few other algae, have resulted in a general acceptance of the idea that evidence from fossil studies is important for a discussion of phylogenetic relationships within relatively small taxa, but is of less use for major discussions of the overall evolutionary trends within a phlyum, or for speculations on the relationship between different phyla. However, the fossil evidence[7, 28, 29, 79] suggests that the dominant kind of pre-Cambrian life was blue-green algal-like, thus supporting the idea (usually inferred from studies of living specimens) that blue-green algae are more primitive than the others. It is to be hoped that future palaeobotanical investigations will yield other evidence of comparable general importance to the question of algal evolution (possibly, analysis of existing data might suggest as much). Until more fossil evidence is available, discussions of algal evolution must continue to be based on inferences from living specimens and their consequent speculative nature recognised.

Two terms commonly used in phylogenetic discussions are *primitive* and *advanced*. Although it is not easy to define these terms exactly, it is necessary to explain briefly what is meant by them. An organism is more advanced than another if at some stage in its evolution it passed through a stage when it resembled the more primitive form. This explanation is clearly limited to those organisms which are related phylogenetically, and critical use of these terms is probably best limited to such forms. That is, their more general use for discussions of unrelated forms is difficult to justify since the terms are then incapable of precise definition.

EVOLUTION WITHIN THE MAJOR PHYLA OF ALGAE

Although a detailed discussion of the subject can only be made by discussing each phylum separately, a few general comments are relevant. Three criteria are commonly used to establish whether a particular type of alga is evolutionarily more *advanced* than another: 1. *Vegetative form*. The developments of particular importance appear to be changes from unicellular to colonial, to filamentous and to parenchymatous. Within multicellular forms such features as method of growth (diffuse, intercalary or apical)

and structural differentiation of the thallus are also emphasised.
2. *Specialisation of the sexual process.* The trend usually emphasised is that from an isogamous through anisogamous to an
oogamous method of sexual reproduction. Detailed increases in
complexity of the reproductive organs are also thought to be
important. 3. *Life-history.* A life-history in which the vegetative
plant is haploid and in which the zygote represents the only
diploid stage is generally thought to be a more primitive condition
than one in which the vegetative plant is diploid and the gametes
represent the sole haploid stage. Isomorphic and heteromorphic
alternation of generations are generally thought to be intermediate
between these two extremes.

Although a more detailed analysis of phylogenetic relationships
would require a more comprehensive discussion of criteria, the
above summary outlines the basic reasoning underlying all
approaches to evolution within an algal phylum.

There are two features of these criteria to suggest that their use
as pointers to algal evolution is justified. First, the developments
which are emphasised are major ones. That is, the increased
morphological elaboration is not of a minor kind such as degree
of branching, number and position of laterals, etc. Rather it is a
major change from one kind of structure (e.g. filamentous) to
another (e.g. parenchymatous). Because of this, the problem of
reduction (i.e. the appearance of a morphologically simpler form
from one more elaborate) is not as serious in algae as in other
plant groups. Secondly, the features being considered can be
compared to those of plants known to be higher than algae in
the evolutionary scale. Thus, flowering plants are parenchymatous, they have apical growth, reproduce by an advanced
oogamous sexual process and the vegetative stage is diploid. By
analogy, therefore, algae which possess these four features
(differing in detail, of course) are generally thought to have arisen
later in evolution than those with none of these features.

Although, therefore, degrees of complexity in algal structure
and function can be specified this should not imply that exact
phylogenetic relationship between a more elaborate and a more
simple form can then be easily deduced. Thus although a parenchymatous alga with an oogamous sexual process possibly arose
during evolution at a later stage than did a filamentous form with
an isogamous sexual process, any phylogenetic relationship
between the two particular algae does not immediately follow.
It is over this last problem that there is most controversy.

Some of the problems of phylogeny within the Chlorophyceae, Phaeophyceae and Rhodophyceae are presented below, but before doing so two generally accepted assumptions can be outlined.

The first of these is that algae of each phylum originated from a unicellular form possibly similar to some of the present-day unicellular species. The general hypothesis that unicellular forms were the starting points of evolution in each phylum is based not only on the occurrence of unicellular species, but also on the observation that multicellular species frequently produce a unicellular stage (a spore or gamete), and that this stage is very similar for diverse species and is also similar to some unicellular species in the particular phylum. Thus, the constant structure of zoospore and spermatozoid in the Phaeophyceae is frequently assumed to indicate that brown algae originated from such a unicellular form, despite the fact that no present-day unicellular species are known.

The second widely accepted feature is the idea of parallel evolution in some divisions. The presence of similar forms in a number of different divisions argues that evolution in these has proceeded in parallel. Although the parallelisms apply most strikingly to the Chlorophyta, Xanthophyta, Chrysophyta and Pyrrophyta, it is also inferred by analogy, for the other divisions, Bacillariophyta, Cryptophyta and the Euglenophyta. In these latter divisions, however, it has not continued for as long so that only unicellular or colonial forms can be identified.

ORIGIN OF THE FILAMENTOUS FORMS IN THE CHLOROPHYCEAE

There are two common hypotheses for the origin of the filamentous habit in the Chlorophyceae. The first suggests an origin from unicellular types, whereas the second suggests that they arose from palmelloid forms. The former is supported by Fritsch[90] who presents as 'evidence' the germination of, for example, a Ulotrichacean zoospore, which, according to this hypothesis, reflects what happened during evolution. Other authorities,[107, 222] however, suggest that filamentous forms arose from palmelloid forms and cite as 'evidence' the existence of palmelloid forms in which the cells are sometimes arranged in linear series.

If one assumes that filamentous forms originated from unicellular ancestors it is unknown whether the latter were motile or non-motile. One hypothesis suggests that non-motile forms evolved from motile and that the filamentous forms arose from the former. Another suggests evolution of all filamentous forms from the motile unicells, with the development of non-motile unicellular (and colonial) species as a 'side-branch' outside this main line of evolution. A third kind suggests that the series of filamentous (and parenchymatous) forms with uninucleate cells arose from the motile unicells, whereas the multinucleate series of forms arose from the Chlorococcales (see below). It is probable that all possible methods actually happened during evolution.

Despite such controversy it is generally accepted that the colonial forms represent a 'blind alley' of evolution unrelated to the evolution of the filamentous thallus. Similarly, it is also accepted that the Oedogoniales and Zygnematales diverged from the main line of evolution (as shown by the Ulotrichales, Chaetopherales and Ulvales) early on, and separate classes (or subclasses) are sometimes suggested for these two groups.

SEPARATION OF MULTINUCLEATE FROM UNINUCLEATE FORMS IN THE CHLOROPHYCEAE

One important problem is whether the series of multinucleate forms (Cladophorales, Acrosiphonales, Siphonocladales, Dasycladales and Caulerpales) are justifiably separated from filamentous forms with uninucleate cells. Related to this is the question of whether the two forms arose from different unicellular types; the uninucleate from the motile unicellular forms of the Volvocales, and the multinucleate from the non-motile Chlorococcales. An alternative to this is the evolution of the Siphonocladales (and subsequently the Caulerpales) from a Ulotrichaceous ancestor rather than from the Chlorococcales. It is difficult to justify the use of a single feature such as number of nuclei as a basis for a major division of green algae into two series. This is particularly true when one is inconsistent; for example, when less importance is attached to it in the Sphaeropleales. Its validity is increased considerably when it can be supported by other features, notably the detailed structure and composition of the cell wall.

PROBLEMS IN THE PHAEOPHYCEAE

Apart from the proposed, but as yet unidentified, unicellular ancestor, the phylogenetic relationships between forms with the three various kinds of life-histories cannot be established. In particular, the problem of how much the three sub-groups represent diverging lines of development and how much certain forms (particulary Laminariales and Fucales) have arisen from relatively elaborate forms of other subclasses?

Electron microscopic studies by Manton and her co-workers[155, 156, 159] have introduced a new approach to speculations on the evolution of brown algae. After examining spermatozoids of *Fucus* and several species of the Cystoseiraceae, Manton concluded that 'the evidence is interpreted as giving strong support to the view that *Fucus* is not primitive but highly specialised and that *Cystoseira* has remained closer to an ancestral condition in a number of different characters'. Although only a few species have so far been examined, this approach may be of general importance in establishing more detailed phylogenetic relationships in the Phaeophyceae.

PROBLEMS IN THE RHODOPHYCEAE

Despite the existence of a few unicellular forms, the origin of the *Rhodophyceae* presents as many problems as does that of the Phaeophyceae, since the unicellular forms are generally assumed to be reduced rather than primitive. Thus, if present-day forms did arise from unicellular Rhodophycean ancestors the latter are unknown. Although the Bangiophycidae are less elaborate than the Florideophycidae it is not generally assumed that the latter originated in the former, rather that they represent divergent lines of evolution from a common ancestral stock. Within each subclass, it is not possible to propose any clearly acceptable scheme of evolution.

The above paragraphs have attempted to present and summarise some of the unsolved problems of algal evolution in such a way as to clarify the questions rather than to suggest answers. The latter too often depend on the opinion of the particular authority to permit adequate treatment here.[23]

RELATIONSHIP BETWEEN DIFFERENT ALGAL PHYLA

In this book, algae have been divided into 10 phyla, each of equal status. That is, any phylogenetic relationship between certain phyla is not emphasised. In the past, authorities wishing to indicate that two or more groups are more closely related to each other than to the remaining algae have generally included them as distinct classes within a single phylum. In particular, several authorities include the Xanthophyceae, Bacillariophyceae and Chrysophyceae in the phylum Chrysophyta.[90, 187] However, the differences in flagellation, the absence of chlorophyll c and fucoxanthin from the Xanthophyceae and certain differences in cell structure argue against this. Differences of pigmentation and cell structure also argue against inclusion of the Dinophyceae and Cryptophyceae in the phylum Pyrrophyta.

Although earlier attempts to suggest affinities between certain classes are less frequently included in taxonomic groupings, there remains the idea that such affinities exist.

The more probable affinities are as follows:

1. On the basis of absence of flagella and presence of biloproteins the Myxophyceae and Rhodophyceae may be closely related.

2. Because chlorophyll b is present in the Chlorophyceae and Euglenophyceae, some kind of relationship between these can be suggested.

3. The presence of chlorophyll c (together with some common carotenoids) argues for an affinity between the Chrysophyceae, Bacillariophyceae, Cryptophyceae, Dinophyceae and Phaeophyceae.

The most recent (and most exciting) attempt to construct a classification scheme which will emphasize the possible relationships is that of Christensen.[26, 27]

On the basis of the difference between procaryotic and eucaryotic cells, he makes the primary classification of algae into *Procaryota* and *Eucaryota*. These appear to be excellent terms, although they are best considered as a primary classification of all cellular organisms and not only algae. Eucaryota are then divided into *Aconta* (no flagellate stages; that is, the Rhodophyta) and *Contophora* (with flagellate stages). He further divides the Contophora into the *Chlorophyta* (in which chlorophylls predominate and in which chlorophyll b is present) and *Chromophyta* (carotenoids predominating and from which chlorophyll b is absent).

Although the importance of Christensen's scheme is recognised, it is not adopted in this book because its various groupings are made by emphasising a *single* feature and ignoring many others. Thus, although some aspects of pigmentation support the scheme, others (notably the carotenoids)[102] do not agree completely with a separation of Chlorophyta from Chromophyta. Similarly, details of nuclear structure (which some authorities would consider to be of fundamental importance) do not support many of the central ideas of the scheme.

However, further investigation may provide additional support for the scheme, and this encouragement to further work is one of its greatest values. Another is the stimulating way Christensen has presented a schematic view of algal evolution (fig. 31). Essentially the scheme consists of two parts: first, the central more or less vertical line, together with its main bifurcation, and secondly, the divergent side branches. The former represents a basic evolution of a few fundamental features of cell structure and biochemistry, whereas the latter represents increasing elaboration of

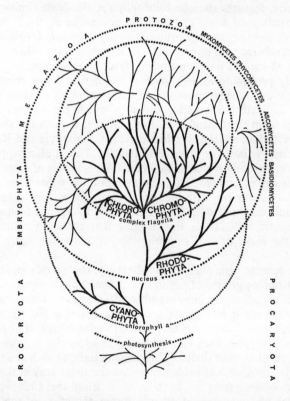

Fig. 31 An outline of Christensen's views on the supposed relationship between the major groups of algae and other organisms. (Reproduced from ref. 27)

morphology and reproduction within each phylum. The first change of the former kind is the development of a eucaryotic type of cell architecture, this is then followed by the appearance of flagella and finally (for our purpose here) the divergence into those with chlorophyll *b* and those without.

As with all attempts to describe the evolution of any group of organisms with an 'evolutionary-tree' diagram, the present one provokes some disagreement and controversy. However, it has two important features which argue in favour of it:

1. The major differences which separate the various algal phyla from one another argue against a *completely monophyletic* origin of them as divergent lines from a common ancestral stock. Similarly, features such as chlorophyll *a*, the basic similarity of cell structure of many of them, the possession of '9+2' flagella by many have argued against a *completely polyphyletic* origin. The present scheme with its central evolving axis and the side branches arising at various points reconciles these two extremes.

2. By proposing an evolution of eucaryotic forms from procaryotic, the scheme promotes an examination of features which might indicate the possible order in which the various phyla diverge from the central axis. For example, the scheme shows a relatively close relationship between the Cyanophyta and Rhodophyta on the basis of absence of flagella. This is also supported by the biloprotein pigments but there is no evidence of a similarity in their cell structure. The Pyrrophyta, on the other hand, have no obvious affinities with blue-green algae but certain features of their nucleus show a striking resemblance to the nucleoplasm of procaryotic cells. Does this suggest that this group diverged from the main axis of evolution earlier than did many other phyla?

The scheme also outlines briefly the probable relationship of algae to other groups of organisms. *Bacteria* are procaryotic and possess a sufficiently large number of features in common with blue-green algae to suggest a common ancestor for these two groups.[71] The photosynthetic pigments and mechanism of photosynthesis suggest that eucaryotic algae arose from blue-green algae (albeit simpler than any in existence today) rather than from bacteria. Biochemical evidence is also the main reason for supposing *higher plants* to have evolved from the Chlorophyta, although the question of where, during evolution of the green algae, the higher plants 'branched off' is not clear.

Clearly the scheme cannot present definite answers to these

diverse problems. However, its stimulation of further investigations, even when they might bring about its downfall, is to be desired. Similarly it serves as a basis for argument and reasoned speculation, and as such should be welcome. It is to be hoped that the present small book goes some way towards achieving these same ends.

REFERENCES

1 ABBAS, A. and GODWARD, M. B. E. (1963). *J. Linn. Soc.* (*Bot.*). **58**, 499–507

2 AFZELIUS, B. A. (1963). *J. Cell Biol.* **19**, 229–38

3 ALLEN, M. B. (1959). *Arch. Mikrobiol.* **32**, 270–7

4 ALLEN, M. B., FRIES, L., GOODWIN, T. W. and THOMAS, D. M. (1964). *J. Gen. Microbiol.* **34**, 259–67

5 ANDERSON, E. (1962). *J. Protozool.* **9**, 380–95

6 ARCHER, A. and BURROWS, E. M. (1960). *Brit. Phycol. Bull.* **2**, 31–3

7 BARGHOORN, E. S. and TYLER, S. A. (1965). *Science*, **147**, 563–77

8 BELCHER, J. H. and FOGG, G. E. (1955). *New Phytol.* **54**, 81–3

9 BEN-SHAUL, Y., SCHIFF, J. A. and EPSTEIN, H. T. (1964). *Plant Physiol.* **39**, 231–40

10 BIDWELL, R. G. S. (1957). *Can. J. Bot.* **35**, 945–50

11 BLACKMAN, F. F. (1900). *Ann. Bot. Lond.*, **14**, 647–88

12 BOGORAD, L. (1962). In *The physiology and biochemistry of algae* (ed. Lewin, R. A.) pp. 385–408. New York and London

13 BOUCK, G. B. (1962). *J. Cell Biol.* **12**, 553–69

14 BOURRELLY, P. (1957). *Rev. algol. Mem. Hors. Ser.* **1**, 1–412

15 BRADFIELD, J. R. G. (1955). In *Sympos. Soc. Exptl. Biol.* **9**, 306–34

16 BRAWERMAN, G. (1962). *Biochim. Biophys. Acta.* **61**, 313–15

17 BRODY, M. and VATTER, A. E. (1962) (1959). *J. Biophys. Biochem. Cytol.* **5**, 289–94

18 BROKAW, C. J. (1962). In *The physiology and biochemistry of algae* (ed. Lewin, R. A.) pp. 595–601. New York and London

19 CHADEFAUD, M. (1960). In *Traité de botanique systématique*. Tome 1 (eds. Chadefaud, M. and Emberger, L.). Paris

20 CHAPMAN, D. J. and CHAPMAN, V. J. (1961). *Ann. Bot. Lond. N.S.* **25**, 547–61

21 CHAPMAN, J. A. and SALTON, M. R. J. (1962). *Arch. Mikrobiol.* **44**, 311–22

22 CHAPMAN, V. J. (1954). *Bull. Torrey Bot. Club.* **81**, 76–82

23 CHAPMAN, V. J. (1962). *The Algae.* London

24 CHEN, Y. T. (1950). *Quart. J. Microscop. Sci.* **91**, 279–308
25 CHIHARA, M. 9 (1960). *J. Jap. Bot.* **35**, 1–00
26 CHRISTENSEN, T. (1962). *Botanik. Systematisk Botanik*, Bd. **11**. No. 2: *Alger*. København
27 CHRISTENSEN, T. (1964). In *Algae and man* (ed. Jackson, D. F.). pp. 59–64. New York
28 CLOUD, P. E., Jr. (1965). *Science*, **148**, 27–35
29 CLOUD, P. E., Jr. and HAGEN, H. (1965). *Proc. Natl. Acad. Sci. U.S.* **54**, 1–8
30 COHN, F. (1871/2). *Jb. Schles. Ges Vaterl. Kult.* **49**, 83
31 COLE, K. and AKINTOBI, S. (1963). *Can. J. Bot.* **41**, 661–8
32 COLEMAN, A. W. (1959). *J. Protozool.* **6**, 249–64
33 COLEMAN, A. W. (1961). *Amer. J. Bot.* **48**, 542
34 COLEMAN, A. W. (1963). *J. Protozool.* **10**, 141–8
35 COSTERTON, J. W. F., MURRAY, R. G. E. and ROBINOW, C. F. (1961). *Can. J. Microbiol.* **7**, 329–39
36 CRESPI, H. J., MANDEVILLE, S. E. and KATZ, J. J. (1962). *Biochem. Biophys. Res. Comm.* **9**, 569–73
37 CRONSHAW, J., MYERS, A. and PRESTON, R. D. (1958). *Biochem. Biophys. Acta.* **27**, 89–103
38 DANGEARD, P. (1954). In VIII *Congr. Int. Bot. Paris.* p. 76
39 DAVIS, J. S. (1964). *Bot. Gazette.* **125**, 129–31
40 DAWES, C. J., SCOTT, F. M. and BOWLER, E. (1961). *Amer. J. Bot.* **48**, 925–34
41 DENFFER, D. von. (1949). *Arch. Mikrobiol.* **14**, 159–202
42 DESIKACHARY, T. V. (1959). *Cyanophyta.* I.C.A.R. New Delhi
43 DESIKACHARY, T. V. and SUNDARALINGAM, V. S. (1962). *Phycologia*, **2**, 9–16
44 DIXON, P. S. (1959). *Botaniska Notiser.* **112**, 339–52
45 DIXON, P. S. (1960). *Botaniska Notiser.* **113**, 295–319
46 DIXON, P. S. (1961). *Bot. Marina.* **3**, 1–16
47 DIXON, P. S. (1962). *Botaniska Notiser.* **115**, 245–60
48 DIXON, P. S. (1963). In *Proc. 4th Int. Seaweed Sympos.* pp. 71–7. Oxford.
49 DIXON, P. S. (1963). *Taxon.* **12**, 108–10
50 DIXON, P. S. (1963). In *Systematics Assn. Publ. no. 5.* pp. 51–62. London
51 DIXON, P. S. (1963). *Ann. Bot. Lond. N.S.* **27**, 353–5
52 DIXON, P. S. (1964). *Botaniska Notiser.* **117**, 56–78
53 DIXON, P. S. (1964). *Botaniska Notiser.* **117**, 279–84
54 DIXON, P. S. (1964). *Nature Lond.* **201**, 519–20
55 DODGE, J. D. (1963). *Arch. Protistenk.* **106**, 442–52
56 DODGE, J. D. (1963). *Arch. Mikrobiol.* **45**, 46–57
57 DODGE, J. D. (1964). *Arch. Mikrobiol.* **48**, 66–80
58 DODGE, J. D. and GODWARD, M. B. E. (1961). *Brit. Phycol. Bull.* **2**, 102–3
59 DOUGHERTY, E. C. and ALLEN, M. B. (1960). In *Comparative biochemistry of photoreactive systems.* (ed. Allen, M. B.) pp. 129–44. New York and London
60 DREW, K. M. (1951). In *Manual of phycology.* (ed. Smith, G. M.) pp. 167–91. Waltham
61 DREW, K. M. (1954). *Ann. Bot. Lond. N.S.* **18**, 183–211

62 DREW, K. M. (1955). *Biol. Rev.* **30**, 343–90
63 DREW, K. M. (1956). *Bot. Rev.* **22**, 553–611
64 DREWS, G. and MEYERS, M. (1964). *Arch. Mikrobiol.* **48**, 259–67
65 DREWS, G. and NIKLOWITZ, W. (1956). *Arch. Mikrobiol.* **25**, 333–51
66 DROUET, F. (1962). *Proc. Acad. Nat. Sci. Philadelphia.* **114**, 191–205
67 DROUET, F. (1963). *Proc. Acad. Nat. Sci. Philadelphia.* **115**, 261–77
68 DROUET, F. and DAILY, W. A. (1956). *Butler Univ. Bot. Studies.* **12**, 1–218
69 ECHLIN, P. (1963). *J. Cell. Biol.* **17**, 212–15
70 ECHLIN, P. (1964). *Arch. Microbiol.* **49**, 267–74
71 ECHLIN, P. and MORRIS, I. (1965). *Biol. Rev.* **40**, 143–87
72 EGEROD, L. E. (1952). *Univ. Calif. Publ. Bot.* **25**, 325–423
73 EPSTEIN, H. T. and SCHIFF, J. A. (1961). *J. Protozool.* **8**, 427–32
74 EPSTEIN, H. T., DE LA TOUR, E. B. and SCHIFF, J. A. (1960). *Nature Lond.* **185**, 825–6
75 EVITT, R. (1963). *Proc. Natl. Acad. Sci. U.S.* **49**, 158–64
76 FAN, K-C. (1959). *Bull. Torrey Bot. Club.* **86**, 1–12
77 FAN, K-C. (1960). *Nova Hedwigia.* **1**, 305–7
78 FAURÉ-FREMIET, E. and ROUILLER, C. (1957). *C.R. Acad. Sci. Paris.* **244**, 2655–7
79 FISCHER, A. C. (1965). *Proc. Natl. Acad. Sci. U.S.* **53**, 1205–15
80 FOGG, G. E. (1951). *Ann. Bot. Lond. N.S.* **15**, 23–36
81 FOGG, G. E. (1956). *Bact. Rev.* **20**, 148–65
82 FOGG, G. E. (1959). In *Sympos. Soc. Exptl. Biol.* **13**, 106–25
83 FOGG, G. E. (1964). In *Algae and man* (ed. Jackson, D. F.). pp. 77–85. New York
84 FOTT, B. (1959). *Algenkunde.* Jena
85 FRANK, H., LEFORT, M. and MARTIN, H. H. (1962). *Biochem. Biophys. Res. Comm.* **7**, 322–5
86 FRANK, H., LEFORT, M. and MARTIN, H. H. (1962). *Z. Naturforsch.* **17b**, 262–73
87 FREUDENTHAL, H. D. (1962). *J. Protozool.* **9**, 45–52
88 FRIEDMANN, I. (1959). *Ann. Bot. Lond. N.S.* **23**, 571–94
89 FRIEDMANN, I. (1960). *Nova Hedwigia.* **1**, 332
90 FRITSCH, F. E. (1935). *The structure and reproduction of the Algae*, Vol. I. Cambridge
91 FRITSCH, F. E. (1945). *The structure and reproduction of the Algae*, Vol. II. Cambridge
92 FRITSCH, F. E. (1951). *Proc. Linn. Soc. Lond.* **162**, 194–211
93 GEITLER, L. (1935). *Bot. Rev.* **1**, 149–61
94 GEITLER, L. (1944). *Flora (Jena).* **37**, 300–33
95 GIBBS, S. P. (1960). *J. Ultrastruct. Res.* **4**, 127–48
96 GIBBS, S. P. (1962). *J. Cell Biol.* **15**, 343–61
97 GIBOR, A. and GRANICK, S. (1962). *J. Cell Biol.* **15**, 599–603
98 GIESBRECHT, P. (1959). *Zbl. Bakt. (1. Abt. Orig.),* **176**, 413–31
99 GIESBRECHT, P. (1962). *Zbl. Bakt. Paras., Infekt., und Hygiene.* **187**, 452–98
100 GODWARD, M. B. E. (1966). *Chromosomes of the Algae.* London
101 GOODWIN, T. W. (1957). *J. Gen. Microbiol.* **17**, 467–73
102 GOODWIN, T. W. (1965). In *Plant pigments* (ed. Goodwin, T. W.). New York and London

103 GREEN, D. (1963). *Comp. Biochem. and Physiol.* **9**, 313–16
104 GRELL, K. G. (1956). *Ann. Rev. Microbiol.* **10**, 307–28
105 GROSS, F. and ZEUTHEN, E. (1948). *Proc. Roy. Soc. B.* **135**, 382–9
106 GUILLARD, R. R. L. (1960). *J. Protozool.* **7**, 262–8
107 GUPTA, A. B. and NAIR, G. U. (1962). *Bot. Gazette*, **124**, 144–6
108 HAMMERLING, J. (1953). *Int. Rev. Cytol.* **2**, 475–98
109 HARRIS, K. and BRADLEY, D. E. (1958). *J. Gen. Microbiol.* **18**, 71–83
110 HARTSHORNE, J. N. (1953). *New Phytol.* **52**, 292–7
111 HAUPT, W. (1962/3). *Ber. Deutsch. Bot. Ges.* **75**, 456
112 HENDEY, N. I. (1937). *Discovery Rept. (Cambridge).* **16**, 151–364
113 HENDEY, N. I. (1954). *J. Mar. Biol. Assn. U.K.* **33**, 335–9
114 HENDEY, N. I. (1959). *J. Queckett Micros. Club.* **5**, 147–75
115 HENDEY, N. I. (1964). *An introductory account of the smaller algae of British coastal waters.* Part V. *Bacillariophyceae* (Diatoms). Fish. Invest. Lond. Ser. 4. H.M.S.O. London
116 HENDEY, N. I., CUSHING, D. H. and RIPLEY, G. W. (1954). *J. Roy. Micros. Soc.* **74**, 22–34
117 HERNDON, W. (1958). *Amer. J. Bot.* **45**, 298–308
118 HOFFMAN, L. R. (1963). *Amer. J. Bot.* **50**, 630ff.
119 HOFFMAN, L. R. and MANTON, I. (1962). *J. Exptl. Bot.* **13**, 443–9
120 HOFFMAN, L. R. and MANTON, I. (1963). *Amer. J. Bot.* **50**, 455–63
121 HOPWOOD, D. A. and GLAUERT, A. M. (1960). *J. Biophys. Biochem. Cytol.* **8**, 813–23
122 JAHN, T. L. (1946). *Quart. Rev. Biol.* **21**, 246–74
123 JAHN, T. L. (1951). In *Manual of phycology* (ed. Smith, G. M.). pp. 69–81. Waltham
124 JAHN, T. L., HARMON, W. M. and LANDMAN, M. (1963). *J. Protozool.* **10**, 358–63
125 JAROSCH, R. (1958). *Protoplasma.* **50**, 277–89
126 JAROSCH, R. (1960). *Phyton (Buenos Aires).* **15**, 43–66
127 JAROSCH, R. (1962). In *The physiology and biochemistry of algae* (ed. Lewin, R. A.). pp. 573–81. New York and London
128 JONSSON, S. (1959). *C.R. Acad. Sci. Paris.* **248**, 835–7
129 JONSSON, S. (1959). *C.R. Acad. Sci. Paris.* **248**, 1565–7
130 JORDE, I. (1933). *Nytt Mag. Naturv.* **73**, 1–19
131 KAMIYA, N. (1959). In *Protoplasmatologia*, Vol. III. Part 3a. (eds. Heilbrunn, L. V. and Weber, F.). Vienna
132 KIRK, J. T. O. (1962). *Biochim. Biophys. Acta.* **56**, 139–51
133 KREGER, D. R. (1962). In *The physiology and biochemistry of algae* (ed. Lewin, R. A.). pp. 315–35. New York and London
134 KUMAR, H. D. (1962). *Nature, Lond.* **196**, 1121–2
135 KYLIN, H. (1937). In *Handbuch der Pflanzenanatomie*. Abt. 11. Bd. 6 T.2. (ed. Linsbauer, K.). Berlin
136 KYLIN, H. (1956). *Die Gattungen der Rhodophyceen.* Lund.
137 LANG, N. J. (1963). *J. Protozool.* **10**, 333–9
138 LAZAROFF, N. and VISHNIAC, W. (1962). *J. Gen. Microbiol.* **28**, 203–210
139 LEEDALE, G. F. (1958). *Arch. Mikrobiol.* **32**, 32–64
140 LEEDALE, G. F. (1962). *Arch. Mikrobiol.* **42**, 237–45
141 LEEDALE, G. F. (1964). *Brit. Phycol. Bull.* **2**, 291–306
142 LEVIN, E. Y. and BLOCK, K. (1964). *Nature, Lond.* **202**, 90–1

143 LEWIN, J. C. (1957). *Can. J. Microbiol.* **3**, 427–33

144 LEWIN, J. C. (1962). In *The physiology and biochemistry of algae* (ed. Lewin, R. A.). pp. 445–55. New York and London

145 LEWIN, J. C. (1962). In *The physiology and biochemistry of algae* (ed. Lewin, R. A.). pp. 457–65. New York and London

146 LEWIN, J. C. and GUILLARD, R. R. L. (1963). *Ann. Rev. Microbiol.* **17**, 373–414

147 LEWIN, J. C., LEWIN, R. A. and PHILLPOTT, D. E. (1958). *J. Gen. Microbiol.* **18**, 418–26

148 LUND, J. W. G. (1959). *Brit. Phycol. Bull.* **1**, 1–17

149 LUND, J. W. G. (1960). *New Phytol.* **59**, 349–60

150 LUND, J. W. G. (1962). In *Sympos. Soc. Gen. Microbiol.* XII. pp. 68–110. Cambridge

151 LYMAN, H., EPSTEIN, H. T. and SCHIFF, J. A. (1961). *Biochim. Biophys. Acta.* **50**, 301–9

152 MAGNE, F. (1960). *Cahiers Biol. Mar.* I, 407–20

153 MAGNE, F. (1960). *C.R. Acad. Sci. Paris.* **250**, 2742–4

154 MANTON, I. (1955). In *Cellular mechanisms in differentiation and growth.* (ed. Rudnick, D.). pp. 61–72. Princeton

155 MANTON, I. (1957). *J. Exptl. Bot.* **8**, 294

156 MANTON, I. (1959). *J. Exptl. Bot.* **10**, 448–61

157 MANTON, I. (1959). *J. Mar. Biol. Assn. U.K.* **38**, 319ff.

158 MANTON, I. (1965). In *Advances in Botanical Research*, Vol. II. (ed. Preston, R. D.)

159 MANTON, I. and CLARKE, B. (1956). *J. Exptl. Bot.* **7**, 416–32

160 MANTON, I. and LEEDALE, G. F. (1961). *J. Mar. Biol. Assn. U.K.* **41**, 145–55

161 MANTON, I. and LEEDALE, G. F. (1961). *J. Mar. Biol. Assn. U.K.* **41**, 519–26

162 MANTON, I. and LEEDALE, G. F. (1961). *Phycologia.* **1**, 37ff.

163 MANTON, I. and LEEDALE, G. F. (1963). *Arch. Mikrobiol.* **45**, 285–303

164 MANTON, I. and LEEDALE, G. F. (1963). *Arch. Mikrobiol.* **47**, 115–36

165 MANTON, I. and PARKE, M. (1960). *J. Mar. Biol. Assn. U.K.* **39**, 275–98

166 MANTON, I. and PARKE, M. (1962). *J. Mar. Biol. Assn. U.K.* **42**, 565–78

167 MANTON, I., OATES, K. and PARKE, M. (1963). *J. Mar. Biol. Assn. U.K.* **43**, 225–38

168 MANTON, I., RAYNS, D. G. and ETTL, H. and PARKE, M. (1965). *J. Mar. Biol. Assn. U.K.* **45**, 241–55

169 MATVIYENKO, O. M. (1962). *Ukrain. Bot. Zhur.* **19**, 45–51

170 McREYNOLDS, J. S. (1961). *Bull. Torrey Bot. Club.* **88**, 397

171 MEEUSE, B. J. D. (1963). In *The physiology and biochemistry of algae* (ed. Lewin, R. A.). pp. 289–313. New York and London

172 MERCER, F. V., BOGORAD, L. and MULLENS, R. (1962). *J. Cell Biol.* **13**, 393–403

173 MILLER, W. H. (1958). *Ann. N.Y. Acad. Sci.* **74**, 204–9

174 MILLER, J. D. A. (1962). In *The physiology and biochemistry of algae* (ed. Lewin, R. A.). pp. 357–70. New York and London

175 MYERS, J. and PRESTON, R. D. (1959). *Proc. Roy. Soc. B.* **150**, 447–55

176 MYERS, J. and PRESTON, R. D. (1959). *Proc. Roy. Soc. B.* **150**, 456–9

177 MYERS, J., PRESTON, R. D. and RIPLEY, G. W. (1956). *Proc. Roy. Soc. B.* **144**, 450–9

178 NAKAYAMA, T. O. M. (1962). In *The physiology and biochemistry of algae* (ed. Lewin, R. A.). pp. 409–20. New York and London

179 NAYLOR, M. (1958). *Ann. Bot. Lond. N.S.* **22**, 205–17

180 NICOLAI, E. and PRESTON, R. D. (1952). *Proc. Roy. Soc. B.* **140**, 244–74

181 NICOLAI, E. and PRESTON, R. D. (1959). *Proc. Roy. Soc. B.* **151**, 244–55

182 O'HEOCHA, C. (1962). In *The physiology and biochemistry of algae* (ed. Lewin, R. A.). pp. 421–35. New York and London

183 PAASCHE, E. (1963). *Physiol. Plant.* **16**, 186–200

184 PAASCHE, E. (1964). *Physiol. Plant.* Suppl. III. pp. 1–82

185 PANKRATZ, H. S. and BOWEN, C. C. (1963). *Amer. J. Bot.* **50**, 387–99

186 PAPENFUSS, G. F. (1951). *Svensk. Bot. Tids.* **45**, 4–11

187 PAPENFUSS, G. F. (1955). In *A century of progress in the natural sciences.* Calif. Acad. Sci. pp. 115–224. San Francisco

188 PARKE, M. (1949). *J. Mar. Biol. Assn. U.K.* **28**, 255–86

189 PARKE, M. (1961). *Brit. Phycol. Bull.* **2**, 47–55

190 PARKE, M. and ADAMS, I. (1960). *J. Mar. Biol. Assn. U.K.* **39**, 263–74

191 PARKE, M. and DIXON, P. S. (1964). *J. Mar. Biol. Assn. U.K.* **44**, 499–542

191a PARKE, M. and MANTON I. (1965). *J. Mar. Biol. Assn. U.K.* **45**, 525–36

192 PARKE, M. and RAYNS, D. G. (1964). *J. Mar. Biol. Assn. U.K.* **44**, 209–17

193 PARKE, M., LUND, J. W. G. and MANTON, I. (1962). *Arch. Mikrobiol.* **42**, 333–52

194 PARKE, M., MANTON, I. and CLARKE, B. (1958). *J. Mar. Biol. Assn. U.K.* **37**, 209–28

195 PARKE, M., MANTON, I. and CLARKE, B. (1959). *J. Mar. Biol. Assn. U.K.* **38**, 169–88

196 PARSONS, T. R., STEPHENS, K. and STRICKLAND, J. D. H. (1961). *J. Fisheries Res. Board Can.* **18**, 1001

197 PATRICK, R. (1954). In AAAS Sympos. Vol. *Sex in micro-organisms.* pp. 82–99

198 PIA, J. (1927). In *Handbuch der Paläobotanik.* Vol. I. (ed. Hirmer, M.). Munich

199 PORTER, K. R. (1955/6). *Harvey Lects.* **51**, 175–228

200 PORTER, K. R. and BRUNI, C. (1960). *Anat. Rec.* **136**, 260–1

201 PRINGSHEIM, E. G. (1949). *Bact. Revs.* **13**, 47–98

202 PROSCHKINA-LAVRENKO, A. I. (1960). *Byull. Mosk. Obshchestva Ispyatelei Prirody Ord. Biol.* **65**, 52

203 RANDHAWA, M. S. (1959). Zygnemaceae. I.C.A.R. New Delhi

204 RIS, H. and SINGH, R. N. (1961). *J. Biophys. Biochem. Cytol.* **9**, 63–80

205 ROBERTS, M. (1964). In *Algae and man.* (ed. Jackson, D. F.). pp. 65–76. New York

206 ROSS, M. M. (1959). *Aust. J. Bot.* **7**, 1–11

207 ROUILLER, C. and FAURÉ-FREMIET, E. (1958). *Exptl. Cell Res.* **14**, 47–67

208 ROUND, F. E. (1963). *Brit. Phycol. Bull.* **2**, 224–35

209 ROUND, F. E. (1965). *The biology of the Algae*. London
210 ROY, K. (1938). *Rev. Algol.* **11**, 101
211 SAGER, R. (1965). *Scientific American.* **212**, 70–9
212 SAGER, R. and PALDE, G. E. (1959) (1957). *J. Biophys. Biochem. Cytol.* **3**, 463–87
213 SARJEANT, W. A. S. (1961). *Grana Palynologia.* **2**, 101
214 SARMA, Y. S. R. K. (1962). *Cytologia.* **27**, 72–8
215 SCHIFF, J. A., LYMAN, H. and EPSTEIN, H. T. (1961). *Biochim. Biophys. Acta.* **50**, 310–18
216 SCHIFF, J. A., LYMAN, H. and EPSTEIN, H. T. (1961). *Biochim. Biophys. Acta.* **51**, 340–6
217 SCHMID, G. (1923). *Jahrb. wiss. Bot.* **62**, 328–419
218 SCHMITZ, F. (1889). *Flora, Jena.* **72**, 435–56
219 SINGH, R. N. and SINHA, R. (1965). *Nature, Lond.* **207**, 782–3
220 SMALL, J. (1950). *Ann. Bot. Lond. N.S.* **14**, 91–113
221 SMITH, G. M. (1933). *The freshwater algae of the United States.* New York
222 SMITH, G. M. (1955). *Cryptogamic botany.* Vol. I. 2nd edn. London
223 SOMA, S. (1960). *J. Fac. Sci. Univ. Tokyo.* Sect. III. **7**, 535–42
224 STANIER, R. Y. and VAN NIEL, C. B. (1941). *J. Bact.* **42**, 437–66
225 STANIER, R. Y. and VAN NIEL, C. B. (1962). *Arch. Mikrobiol.* **42**, 17–35
226 STEELE, J. H. and YENTSCH, C. S. (1960). *J. Mar. Biol. Assn. U.K.* **39**, 217–26
227 STEEMAN NIELSEN, E. (1963). *Physiol. Plant.* **16**, 466–9
228 STEIN, J. R. (1958). *Amer. J. Bot.* **45**, 664
229 STERN, A. I., EPSTEIN, H. T. and SCHIFF, J. A. (1964). *Plant Physiol.* **39**, 226–31
230 STERN, A. I., SCHIFF, J. A. and EPSTEIN, H. T. (1964). *Plant Physiol.* **39**, 220–6
231 STOSCH, H. A. von (1955). *Naturwissenschaften.* **42**, 423
232 STOSCH, H. A. von (1958). *Naturwissenschaften.* **45**, 140–1
233 STOSCH, H. A. von (1958). *Arch. Mikrobiol.* **31**, 274–82
234 STRAIN, H. H. (1951). In *Manual of phycology* (ed. Smith, G. M.). pp. 243–62. Waltham
235 SVEDELIUS, N. (1942). *Blumea,* Suppl. **2**, 72–90
236 TRAINOR, F. R. (1963). *Science.* **142**, 1673–4
237 TRAINOR, F. R. (1963). *Can. J. Bot.* **41**, 967–8
238 UEDO, K. (1958). *Cytologia.* **23**, 56–67
239 WALL, D. (1962). *Geol. Mag.* **99**, 353–62
240 WIESE, L. and JONES, R. F. (1963). *J. Cell. Comp. Physiol.* **61**, 265–74
241 WILDON, D. C. and MERCER, F. V. (1963). *Arch. Mikrobiol.* **47**, 19–31
242 WILBUR, K. M. and WATABE, N. (1963). *Ann. N.Y. Acad. Sci.* **109**, 82–112
243 WOLK, C. P. (1965). *Nature, Lond.* **205**, 201–2
244 WOLKEN, J. J. (1956). *J. Protozool.* **3**, 211–21
245 WOLKEN, J. J. and PALADE, G. E. (1953). *Ann. N.Y. Acad. Sci.* **56**, 873–89
246 ZAHL, P. A. and McLAUGHLIN, J. J. (1959). *J. Protozool.* **6**, 344–52

INDEX

Bold figures refer to line-drawings